Shortwave Listener's

Guide

by
H. Charles Woodruff

Howard W. Sams & Co., Inc.
4300 WEST 62ND ST. INDIANAPOLIS, INDIANA 46268 USA

Preface

Every owner of a shortwave receiving set is familiar with the thrill that comes from hearing a distant station broadcasting from a foreign country. To hundreds of thousands of people the world over, shortwave listening (often referred to as swl) represents the most satisfying, the most worthwhile of all hobbies.

It has been estimated that more than 25 million shortwave receivers are in the hands of the American public, with the number increasing daily. To explore the international shortwave broadcasting bands in a knowledgeable manner, the shortwave listener must have available a list of shortwave stations, their frequencies, and their times of transmission. To keep abreast of the ever-increasing public interest in music, news, and the exchange of cultural ideas from foreign lands, the seventh edition of this book has been completely revised to include the most recent changes in broadcasting schedules. A new format has been adopted in order to present the shortwave schedules and frequency information in a more comprehensive style and in a manner that can be utilized by shortwave enthusiasts the world over. The listings presented in this edition are arranged conveniently in five sections as follows:

Section 1 contains a brief explanation of shortwave propagation in easy-to-understand language, along with forecast tables that the swl'er can use as a guide to the best wavelength bands to be used for listening to stations in various geographic locations. These tables cover from early 1978 through February 1980. Section 1 also contains a do-it-yourself method that will enable the nontechnical swl enthusiast to predict daily ionospheric conditions with a high degree of accuracy.

Section 2 consists of worldwide shortwave broadcasting stations listed alphabetically according to country and city within the country. Important particulars such as geographic location, rf carrier output in kilowatts (kW), interval signal (Int), announcement (Ann), frequency in kilohertz (kHz), and hours of transmission in Coordinated Universal Time (UTC) for each station are given.

Section 3 contains a listing of shortwave broadcasting stations, including the country and city, in numerical order by frequency.

Section 4 contains a listing of shortwave news broadcasts that are transmitted in English and other principal languages. The tabulation is arranged by UTC and alphabetically by country.

Section 5 includes the official names and business addresses of the shortwave stations so that the shortwave listener can send verification reports of transmissions heard. QSL-ing, as it is called, is explained in this section.

The tabulations in this book by no means represent all of the shortwave broadcasting stations in the world. Only those transmitting in English and a few of the other major languages of the world are included. Every effort has been expended to make this edition as useful and up to date as possible. However, the accuracy cannot be guaranteed; carrier frequencies and program scheduling may change without notice. All swl'ers are cordially invited to comment on any additions, deletions, or changes that may be noted, so that these can be incorporated in future editions.

H. Charles Woodruff

Contents

SECTION 5

SECTION 6

Introduction

To pursue the very interesting and stimulating hobby of shortwave listening in an informed manner, the hobbyist must be aware of a few salient facts. These important items are discussed in the following paragraphs. Every effort has been made to simplify the data. If detailed information on a particular subject is desired, it is suggested that a textbook be consulted.

FREQUENCY

All transmission frequencies in this book are expressed in kilohertz (kHz). Receiver dials are often calibrated in megahertz (MHz). To convert kilohertz to megahertz, simply divide by 1000; conversely, to convert megahertz to kilohertz, multiply by 1000. For example, 9100 kHz is 9.1 MHz, and 5 MHz is 5000 kHz. The tuning dials of most shortwave receivers omit fractional numbers. For example, the numerals 9, 10, 15, 20, etc., appearing on the tuning dial stand for 9 MHz, 10 MHz, 15 MHz, and 20 MHz. To determine the location of a station broadcasting on 9.100 megahertz, the operator needs only to mentally divide the space between 9 MHz and 10 MHz, and position the receiver tuning-dial marker one-tenth of that spacing beyond 9 MHz. Some receivers have precision dial calibrations that are expressed directly in kilohertz.

Some receivers may use the terms kilocycles (kc) or megacycles (mc) instead of kilohertz and megahertz. The terms have the same significance; that is, "kilocycles" is the same as "kilohertz" and "megacycles" is the same as "megahertz." Formerly, the term "cycles per second" was used to designate frequencies. The terms kilocycles and megacycles (actually kilocycles per second and megacycles per second) were used to designate 1000 and 1,000,000 cycles per second. The newer term, hertz, was adopted partially because of the fact that the "per second" portion of the previous designation was often omitted (though without the time element the term is meaningless) and partially to honor Heinrich Hertz, considered by many as the father of radio. The term hertz (Hz) means cycles per second; thus, the time is included as part of the term. Likewise, kilohertz means 1000 cycles per second (1000 Hz) and megahertz means 1,000,000 cycles per second (1,000,000 Hz or 1000 kHz).

By international regulations, entered into by most countries, certain groups, or "bands," of radio frequencies have been set aside in the high-frequency radio spectrum for international shortwave broadcasting (Table 1). Most of the world's shortwave broadcasting stations operate within these bands; some, however, operate outside the band limits, usually adjacent to a particular band within 100 or 200 kHz. Occasionally you might hear that a particular station is operating in the "16-meter band," or the "41-meter band," etc. Table 1 lists these designations for the various bands.

Table 1. International Shortwave Broadcasting Bands

Frequency (kHz)	Band (Meters)
2300–2495	120
3200–3400	90
3900–4000	75
4750–5060	60
5950–6200	49
7100–7300	41
9500–9775	31
11,700–11,975	25
15,100–15,450	19
17,700–17,900	16
21,450–21,750	13
25,600–26,100	11

CALL LETTERS

Any listener to conventional radio and television is aware of the call letters assigned to transmitting stations, for example, KFI, Los Angeles; KOA, Denver; WLS, Chicago; WNBC, New York; etc. Some countries have assigned call letters to their high-frequency stations; however, the call letters are rarely used for station identification. Usually the announcer of a shortwave station will merely say, "This is Radio Japan," "This is RSA, Radio South Africa," or "This is the Voice of America."

POWER

The transmitting power listed in Section 2 of this book is expressed in kilowatts (kW). Most international shortwave stations use transmitting equipment with a power of 50 kW or more to ride through the interference and atmospheric noise. This high power does not mean that stations of 5 kW or less cannot be heard. Quite the contrary—amateur shortwave operators have repeatedly disproved this by conversing with fellow "hams" all over the world using considerably less than 1 kW of power. The unpredictability of shortwave listening is what makes the hobby interesting and the end result more rewarding.

PROGRAM TARGET AREAS

Most international shortwave broadcasting stations employ directional antenna arrays to beam radio transmissions to a specific geographical area, such as Europe, North America, Africa, Asia, etc. These programmed areas are indicated in the Section 2 station listings.

In most instances, due to overlap of target areas and because of the nature of shortwave radio wave propagation, you may hear broadcasts that are not specifically beamed to your area. For example, BBC transmissions directed to Africa can often be received in Southeast Asia. However, for best reception and maximum signal strength, choose a transmission that is beamed to your geographical location. Those transmissions not indicated as beamed to a specific area can be assumed to be omnidirectional.

All broadcasts listed in *Shortwave Listener's Guide* are daily transmissions unless overwise noted.

MULTIPLE-FREQUENCY TRANSMISSIONS

To compensate for adverse ionospheric conditions, interference from other radio stations, jamming, and technical malfunctions, stations may use several frequencies beamed to the same general geographical area. When radio interference or low received signal strength is encountered, the listener should try a different listed frequency or tune to the station at another time.

TIME ZONES AND LOCAL TIME

The United States is divided into seven standard time zones, designated as Eastern, Central, Mountain, Pacific, Yukon, Alaska/Hawaii, and Bering. The Canadian provinces are in the first five of these seven zones, plus the Atlantic time zone on the east. In addition, Newfoundland and Labrador advance the clock one-half hour ahead of Atlantic time. The various time zones are shown in Fig. 1. Each time zone is approximately 15 degrees of longitude in width, and all places within a given zone use the time reckoned from the transit of the sun across the standard time meridian of that zone. The time for each zone, starting with the Atlantic time zone and moving westward, is basically reckoned from the 60th, 75th, 90th, 105th, 120th, 135th, 150th, and 160th meridians west of Greenwich, England (prime meridian). The actual division lines separating the various time zones

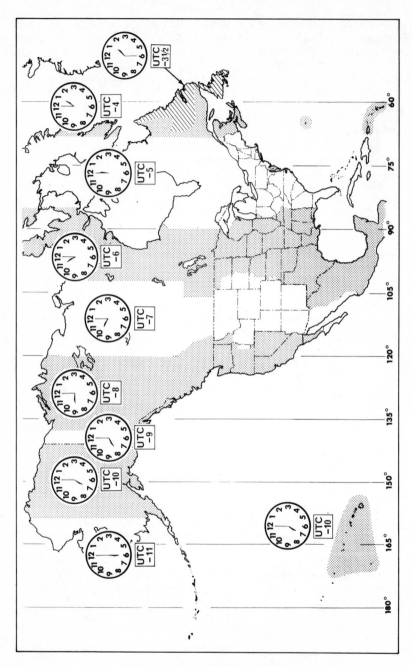

Fig. 1. North American time zones.

wander somewhat to conform with local geographic areas and local convenience.

The times of all events listed in *Shortwave Listener's Guide* are given in Coordinated Universal Time (UTC), which is coordinated through international agreements by the International Time Bureau (BIH, which stands for *Bureau International d l'Heure*, located in Paris, France) so that time signals broadcast from standard frequency and time stations throughout the world (such as WWV in Fort Collins, Colorado) will be in close agreement. The specific hour and minute listed is actually the time in the reference time zone centered around Greenwich, England, and is generally equivalent to the more familiar "Greenwich Mean Time" (GMT). To obtain the local time of the event, it is necessary to add or subtract a given number of hours from the UTC. For example, to obtain the local time in the Eastern time zone, five hours must be subtracted from the time shown; for the Pacific time zone, eight hours must be subtracted.

The standard time differences for principal cities of the United States and Canada are indicated in Table 2. Standard time differences for cities throughout the rest of the world are indicated in Table 3.

Table 2. Standard Time Differences, US and Canadian Cities*

City	Time	City	Time
Akron, OH	1200	Edmonton, Alta	1000
Albuquerque, NM	1000	El Paso, TX	1000
Anchorage, AK	0700	Erie, PA	1200
Atlanta, GA	1200	Evansville, IN	1100
Austin, TX	1100	Fairbanks, AK	0700
Baltimore, MD	1200	Flint, MI	1200
Birmingham, AL	1100	Fort Wayne, IN	1200
Bismarck, ND	1100	Fort Worth, TX	1100
Boise, ID	1000	Frankfort, KY	1200
Boston, MA	1200	Galveston, TX	1100
Buffalo, NY	1200	Gander, Nfld	1330
Butte, MT	1000	Grand Rapids, MI	1200
Charleston, SC	1200	Halifax, N.S.	1300
Charlotte, NC	1200	Hartford, CT	1200
Chattanooga, TN	1200	Helena, MT	1000
Cheyenne, WY	1000	Hilo, HI	0700
Chicago, IL	1100	Honolulu, HI	0700
Cincinnati, OH	1200	Houston, TX	1100
Cleveland, OH	1200	Indianapolis, IN	1200
Colorado Sprs., CO	1000	Jacksonville, FL	1200
Columbus, OH	1200	Juneau, AK	0900
Dallas, TX	1100	Kansas City, MO	1100
Dayton, OH	1200	Knoxville, TN	1200
Denver, CO	1000	Las Vegas, NV	0900
Des Moines, IA	1100	Lexington, KY	1200
Detroit, MI	1200	Lincoln, NE	1100
Duluth, MN	1100	Little Rock, AR	1100
Dutch Harbor, AK	0600	Los Angeles, CA	0900

Table 2. Standard Time Differences, US and Canadian—cont

City	Time	City	Time
Louisville, KY	1200	Richmond, VA	1200
Memphis, TN	1100	Rochester, NY	1200
Miami, FL	1200	Sacramento, CA	0900
Milwaukee, WI	1100	St. Louis, MO	1100
Minneapolis, MN	1100	St. Paul, MN	1100
Mobile, AL	1100	Salt Lake City, UT	1000
Montreal, Que	1200	San Antonio, TX	1100
Nashville, TN	1100	San Diego, CA	0900
Newark, NJ	1200	San Francisco, CA	0900
New Haven, CT	1200	Santa Fe, NM	1000
New Orleans, LA	1100	Savannah, GA	1200
New York, NY	1200	Seattle, WA	0900
Nome, AK	0600	Shreveport, LA	1100
Norfolk, VA	1200	Sioux Falls, SD	1100
Oklahoma City, OK	1100	Spokane, WA	0900
Omaha, NE	1100	Tacoma, WA	0900
Ottawa, Ont	1200	Tampa, FL	1200
Peoria, IL	1100	Toledo, OH	1200
Philadelphia, PA	1200	Topeka, KS	1100
Phoenix, AZ	1000	Toronto, Ont	1200
Pierre, SD	1100	Tucson, AZ	1000
Pittsburgh, PA	1200	Tulsa, OK	1100
Portland, ME	1200	Vancouver, B.C.	0900
Portland, OR	0900	Washington, DC	1200
Providence, RI	1200	Wichita, KS	1100
Quebec, Que	1200	Wilmington, DE	1200
Reno, NV	0900	Winnipeg, Man	1100

* At 1700 hours UTC, the standard time in US and Canadian cities is as listed.

By Federal law, daylight saving time (DST) is observed in the United States from 2:00 a.m. on the last Sunday in April to 2:00 a.m. on the last Sunday in October. (A few states have elected to exempt themselves from the observance of daylight saving time.) Daylight saving time is achieved by advancing the clocks one hour. For example, an event listed for 9:00 p.m. EST would take place at 10:00 p.m. EDT.

Table 3. Standard Time Differences, World Cities*

City	Time	City	Time
Abidjan, Ivory Coast	1700	Bangkok, Thailand	0000
Addis Ababa, Ethiopia	2000	Belfast, Northern Ireland	1800
Adelaide, Australia	0230†	Berlin, Germany	1800
Alexandria, Egypt	1900	Bern, Switzerland	1800
Amsterdam, Netherlands	1800	Bogota, Colombia	1200
Athens, Greece	1900	Bombay, India	2230
Auckland, New Zealand	0500†	Brazzaville, Congo	1800
Baghdad, Iraq	2000	Bremen, Germany	1800

Table 3. Standard Time Differences, World—cont

City	Time	City	Time
Brussels, Belgium	1800	Oslo, Norway	1800
Bucharest, Romania	1900	Panama, Panama	1200
Budapest, Hungary	1800	Paris, France	1800
Buenos Aires, Argentina	1400	Peking, China	0100†
Calcutta, India	2230	Perth, Australia	0100†
Cape Town, So. Africa	1900	Port Moresby, Papua	0300†
Caracas, Venezuela	1300	Prague, Czechoslovakia	1800
Colombo, Sri Lanka	2230	Quito, Equador	1200
Copenhagen, Denmark	1800	Rangoon, Burma	2330
Djakarta, Indonesia	0000	Reykjavik, Iceland	1700
Dublin, Ireland	1800	Rio de Janeiro, Brazil	1400
Geneva, Switzerland	1800	Rome, Italy	1800
Godthab, Greenland .	1400	Saigon, Vietnam	0100†
Guam Island (US)	0300†	Salisbury, Rhodesia	1900
Havana, Cuba	1300	Santiago, Chile	1300
Helsinki, Finland	1900	Seoul, Korea	0200†
Hong Kong	0100†	Shanghai, China	0100†
Istanbul, Turkey	1900	Singapore, Singapore	0030†
Johannesburg, So. Africa	1900	Sofia, Bulgaria	1900
Karachi, Pakistan	2200	Stockholm, Sweden	1800
Kiev, USSR	2000	Sydney, Australia	0300†
Le Havre, France	1800	Tahiti, Fr. Polynesia	0700†
Leningrad, USSR	2000	Taiwan (Nationalist China)	0100†
Lima, Peru	1200	Tashkent, USSR	2300
Lisbon, Portugal	1800	Tehran, Iran	2030
Liverpool, England	1800	Tel Aviv, Israel	1900
London, England	1800	Tokyo, Japan	0200†
Madrid, Spain	1800	Ulan Bator, Mongolia	0100†
Managua, Nicaragua	1100	Valparaiso, Chile	1300
Manila, Philippines	0100†	Vienna, Austria	1800
Mandalay, Burma	2330	Vientiane, Laos	0000
Melbourne, Australia	0300†	Vladivostok, USSR	0300†
Mexico City, Mexico	1100	Warsaw, Poland	1800
Montevideo, Uruguay	1400	Wellington, New Zealand	0500†
Moscow, USSR	2000	Yokohama, Japan	0200†
Omdurman, Sudan	1900	Zurich, Switzerland	1800

* At 1700 hours UTC, the standard time in world cities is as listed.
† Next day.

Notice that in this book times are given in the 24-hour system. In this system, time is expressed by a four-digit number in which the first two digits indicate the hour and the last two digits indicate the minute. The hours are numbered starting with midnight as zero. For example, 12:30 a.m. is 0030, 10:30 a.m. is 1030, 1:00 p.m. is 1300, 6:25 p.m. is 1825, etc. Usually, midnight is written 0000, but occasionally it may appear as 2400.

1

Ionospheric Propagation
and Predictions

Two types of radio-frequency waves are emitted from a shortwave transmitting antenna—the ground wave and the sky wave. The ground wave is of no significance for shortwave reception. The sky wave, however, upon leaving the transmitting antenna travels upward at various angles above the surface of the earth. It would simply continue out into space were it not bent sufficiently to bring it back to the earth. The medium that causes such bending is the ionosphere, a region in the upper atmosphere where free ions and electrons exist in sufficient quantity to cause a change in the refractive index. Ultraviolet radiation from the sun is considered to be responsible for the ionization. For a given intensity of ionization, the amount of refraction becomes less as the frequency of the wave becomes higher. The bending is smaller, therefore, at high frequencies than it is at low frequencies. If the frequency is raised to a high enough value, the bending eventually will become too slight to bring the wave back to earth. At frequencies beyond this point, long-distance shortwave communication becomes impossible.

Because an increase in ionization causes an increase in the maximum frequency of waves that can be bent sufficiently for long-distance communication, it can be seen that slight variations in sun radiation caused by sunspots, solar flares, and other solar disturbances can affect shortwave signal reception.

The amount of ultraviolet radiation reaching the earth varies considerably during a 24-hour period due to the rotation of the earth and because of sunspots and solar flares. Furthermore, the average sunspot number varies over an approximate 11-year cycle. At the time of maximum sunspots, station reception on the 11-meter band is possible over great distances. However, at the time of low sunspot number the 11- and 13-meter bands are virtually useless. Solar activity reached a maximum in 1969 and then declined until it reached its present low in the latter part of 1974. It is expected to remain at this low ebb until 1978, at which time it is predicted to rise slowly.

At times, ionospheric conditions may cause a temporary "signal blackout" from some areas of the earth. Therefore, even though a station might be listed as being "on the air" for a particular time period, ionospheric conditions may prevent the signal from being heard.

During the past several years, communications engineers and scientists have conducted a concentrated worldwide effort to learn more about the mysteries of the ionosphere, its makeup, and its effect on radio transmission and reception. Many satellites and space laboratories have been launched and more than 100 ground stations have been established to study the phenomenon. The results of this continuing study have produced a much greater understanding of electromagnetic wave propagation, and the mass of data compiled by these satellites and ground stations is constantly being fed into computers and data processing machines and used for predicting future propagation conditions. Tables 1-1 through 1-8 have been compiled from such computer programs, and are based on the latest propagation data available. The tables are simple to use. First, locate the chart containing the current month. Second, find your geographic location in the left-hand column titled "Listener's Area," and in the appropriate Coordinated Universal Time (UTC) horizontal line, move to the right to the desired "Broadcasting Station Location" vertical column. Use the following legend to identify your listening area and the areas of the world in which the broadcasting stations are located:

Legend:
  I—North America (East)
  II—North America (West)
  III—Central and South America
  IV—Europe and North Africa
  V—Central and South Africa
  VI—Middle East and South Asia
  VII—East Asia and Far East
  VIII—Australia and New Zealand

The tables show the wavelength band (in meters) that should provide optimum reception in the listener's area for the month and time listed. In some instances, because of slight daily variations in ionospheric density, it may be possible to receive broadcasting stations in the next higher or next lower wavelength bands; however, best reception can be expected on the bands listed.

These tables show the *predicted* wavelength bands that will provide the listener with optimum reception. However, these are long-range predictions and can be subjected to daily fluctuations. The following paragraphs will describe a daily, do-it-yourself method that

Table 1-1. Favorable Shortwave Broadcast Bands (March–April 1978)

Listener's Area*	Time (UTC)	Broadcasting Station Location*							
		I	II	III	IV	V	VI	VII	VIII
I	0000–0400	49	25	31	49	31	31	25	19
	0400–0800	49	49	49	49	49	49	49	31
	0800–1200	49	49	49	49	49	49	49	49
	1200–1600	49	31	25	25	25	25	49	31
	1600–2000	25	25	19	19	19	19	31	19
	2000–2400	25	19	19	25	19	25	25	19
II	0000–0400	31	31	25	49	31	31	19	19
	0400–0800	49	49	49	49	31	25	19	19
	0800–1200	49	49	49	49	41	31	49	31
	1200–1600	31	49	31	31	25	31	49	49
	1600–2000	25	31	19	25	19	19	31	49
	2000–2400	19	19	16	31	19	25	19	19
III	0000–0400	49	25	49	41	31	31	19	19
	0400–0800	49	31	49	49	49	31	31	25
	0800–1200	49	49	49	31	31	25	31	25
	1200–1600	19	31	31	19	19	19	31	25
	1600–2000	19	19	25	19	16	19	25	25
	2000–2400	25	19	19	25	19	25	25	19
IV	0000–0400	49	49	49	49	31	41	31	31
	0400–0800	49	49	49	75	31	31	31	25
	0800–1200	49	49	49	41	19	19	19	19
	1200–1600	25	49	25	31	19	19	19	25
	1600–2000	19	25	19	31	19	25	31	31
	2000–2400	25	25	19	49	25	31	31	31
V	0000–0400	49	31	31	41	49	49	49	25
	0400–0800	49	31	25	25	25	19	19	19
	0800–1200	25	31	16	19	19	16	16	19
	1200–1600	19	25	16	19	25	19	25	31
	1600–2000	19	19	16	31	31	31	31	31
	2000–2400	31	25	25	31	31	49	41	25
VI	0000–0400	31	31	25	49	49	31	25	25
	0400–0800	49	31	31	31	19	25	19	19
	0800–1200	49	31	25	19	19	25	25	31
	1200–1600	25	25	19	19	16	31	31	31
	1600–2000	19	25	19	25	25	41	49	31
	2000–2400	25	25	25	49	49	49	41	31
VII	0000–0400	25	19	19	41	49	25	25	19
	0400–0800	49	25	25	25	31	19	19	16
	0800–1200	49	31	31	19	19	25	25	25
	1200–1600	31	49	31	19	19	31	41	31
	1600–2000	25	31	25	31	31	41	41	41
	2000–2400	25	25	25	41	31	41	41	31
VIII	0000–0400	19	16	19	31	25	31	19	31
	0400–0800	31	25	19	31	25	19	16	25
	0800–1200	49	31	25	25	19	19	19	25
	1200–1600	31	49	25	19	19	25	25	31
	1600–2000	25	49	25	25	25	31	41	49
	2000–2400	25	25	25	31	31	41	41	49

* See legend on page 16.

Table 1-2. Favorable Shortwave Broadcast Bands (May–August 1978)

Listener's Area*	Time (UTC)	Broadcasting Station Location*							
		I	II	III	IV	V	VI	VII	VIII
I	0000–0400	31	25	25	25	25	31	19	19
	0400–0800	49	31	31	31	41	31	31	25
	0800–1200	49	49	31	31	41	31	31	31
	1200–1600	31	31	25	25	19	19	25	31
	1600–2000	25	25	19	19	19	19	19	19
	2000–2400	19	19	19	19	19	19	19	19
II	0000–0400	25	25	25	25	25	25	19	16
	0400–0800	31	31	31	31	31	25	19	25
	0800–1200	49	49	49	31	31	31	31	31
	1200–1600	31	49	31	31	25	25	31	31
	1600–2000	25	31	19	25	19	19	25	31
	2000–2400	19	19	19	25	25	25	19	19
III	0000–0400	31	25	49	31	31	25	19	19
	0400–0800	31	31	49	31	49	25	25	25
	0800–1200	49	49	49	41	31	25	25	25
	1200–1600	19	31	31	19	19	19	25	31
	1600–2000	19	19	25	19	16	19	19	19
	2000–2400	25	19	31	25	25	25	19	19
IV	0000–0400	31	31	31	49	31	31	31	31
	0400–0800	31	31	31	41	31	31	25	25
	0800–1200	31	31	31	31	19	19	19	19
	1200–1600	25	31	19	31	19	16	19	25
	1600–2000	19	25	19	31	19	25	25	31
	2000–2400	25	25	25	31	25	31	31	31
V	0000–0400	49	49	49	41	49	49	49	19
	0400–0800	31	31	25	25	25	19	19	19
	0800–1200	25	31	19	19	25	19	16	25
	1200–1600	19	19	19	16	25	19	25	31
	1600–2000	19	19	19	31	31	31	25	31
	2000–2400	25	31	25	31	49	41	31	31
VI	0000–0400	25	25	25	49	49	31	25	25
	0400–0800	49	31	25	25	19	25	19	19
	0800–1200	25	31	25	16	19	25	19	25
	1200–1600	25	25	19	19	19	31	31	31
	1600–2000	19	25	19	25	31	41	25	49
	2000–2400	25	25	25	31	49	49	31	31
VII	0000–0400	19	19	19	31	41	25	25	16
	0400–0800	25	25	25	25	25	19	31	25
	0800–1200	31	31	25	25	16	25	25	25
	1200–1600	25	31	25	19	19	25	41	31
	1600–2000	25	25	25	31	31	31	49	41
	2000–2400	19	19	19	49	31	31	31	31
VIII	0000–0400	19	19	19	25	31	31	19	31
	0400–0800	25	19	25	31	25	25	16	31
	0800–1200	25	31	25	19	19	19	19	31
	1200–1600	31	31	31	19	19	25	31	31
	1600–2000	25	31	25	25	25	31	41	41
	2000–2400	19	25	25	31	31	41	41	49

* See legend on page 16.

Table 1-3. Favorable Shortwave Broadcast Bands (Sept–Oct 1978)

Listener's Area*	Time (UTC)	Broadcasting Station Location*							
		I	II	III	IV	V	VI	VII	VIII
I	0000–0400	49	25	31	49	31	31	25	19
	0400–0800	49	49	49	49	49	49	49	31
	0800–1200	49	49	49	49	49	49	49	49
	1200–1600	49	31	25	25	25	25	49	31
	1600–2000	25	25	19	19	19	19	31	19
	2000–2400	25	19	19	25	19	25	25	19
II	0000–0400	31	31	25	49	31	31	19	19
	0400–0800	49	49	49	49	31	25	19	19
	0800–1200	49	49	49	49	41	31	49	31
	1200–1600	31	49	31	31	25	31	49	49
	1600–2000	25	31	19	25	19	19	31	49
	2000–2400	19	19	16	31	19	25	19	19
III	0000–0400	49	25	49	41	31	31	19	19
	0400–0800	49	31	49	49	49	31	31	25
	0800–1200	49	49	49	31	31	25	31	25
	1200–1600	19	31	31	19	19	19	31	25
	1600–2000	19	19	25	19	16	19	25	25
	2000–2400	25	19	19	25	19	25	25	19
IV	0000–0400	49	49	49	49	31	41	31	31
	0400–0800	49	49	49	75	31	31	31	25
	0800–1200	49	49	49	41	19	19	19	19
	1200–1600	25	49	25	31	19	19	19	25
	1600–2000	19	25	19	31	19	25	31	31
	2000–2400	25	25	19	49	25	31	31	31
V	0000–0400	49	31	31	41	49	49	49	25
	0400–0800	49	31	25	25	25	19	19	19
	0800–1200	25	31	16	19	19	16	16	19
	1200–1600	19	25	16	19	25	19	25	31
	1600–2000	19	19	16	31	31	31	31	31
	2000–2400	31	25	25	31	31	49	41	25
VI	0000–0400	31	31	25	49	49	31	25	25
	0400–0800	49	31	31	31	19	25	19	19
	0800–1200	49	31	25	19	19	25	25	31
	1200–1600	25	25	19	19	16	31	31	31
	1600–2000	19	25	19	25	25	41	49	31
	2000–2400	25	25	25	49	49	49	41	31
VII	0000–0400	25	19	19	41	49	25	25	19
	0400–0800	49	25	25	25	31	19	19	16
	0800–1200	49	31	31	19	19	25	25	25
	1200–1600	31	49	31	19	19	31	41	31
	1600–2000	25	31	25	31	31	41	41	41
	2000–2400	25	25	25	41	31	41	41	31
VIII	0000–0400	19	16	19	31	25	31	19	31
	0400–0800	31	25	19	31	25	19	16	25
	0800–1200	49	31	25	25	19	19	19	25
	1200–1600	31	49	25	19	19	25	25	31
	1600–2000	25	49	25	25	25	31	41	49
	2000–2400	25	25	25	31	31	41	41	49

* See legend on page 16.

Table 1-4. Favorable Shortwave Broadcast Bands (Nov 1978–Feb 1979)

Listener's Area*	Time (UTC)	Broadcasting Station Location*							
		I	II	III	IV	V	VI	VII	VIII
I	0000–0400	49	31	49	49	31	49	25	25
	0400–0800	49	49	49	75	49	49	31	25
	0800–1200	49	49	49	75	49	49	41	49
	1200–1600	49	31	25	25	25	25	31	31
	1600–2000	25	25	16	19	19	19	31	25
	2000–2400	31	19	19	31	25	25	19	25
II	0000–0400	31	31	31	49	31	25	19	19
	0400–0800	49	49	49	49	49	31	31	31
	0800–1200	49	49	49	75	49	31	49	31
	1200–1600	49	49	31	49	25	31	49	49
	1600–2000	25	49	19	25	19	25	31	31
	2000–2400	25	25	25	25	25	31	25	16
III	0000–0400	49	31	49	49	31	49	19	19
	0400–0800	49	49	49	49	41	49	31	25
	0800–1200	49	49	49	41	31	31	31	25
	1200–1600	19	31	31	19	19	19	31	25
	1600–2000	19	19	25	19	16	19	25	25
	2000–2400	25	19	31	31	25	31	31	19
IV	0000–0400	49	49	49	75	41	41	41	41
	0400–0800	49	49	49	75	31	41	31	31
	0800–1200	49	49	31	49	19	19	19	31
	1200–1600	31	49	25	31	19	19	19	25
	1600–2000	19	31	19	41	25	31	41	31
	2000–2400	31	31	25	49	31	41	49	41
V	0000–0400	49	49	31	41	41	31	41	25
	0400–0800	49	31	31	25	25	19	25	19
	0800–1200	31	31	25	19	25	16	19	25
	1200–1600	19	25	16	19	25	19	25	31
	1600–2000	25	19	19	31	31	31	41	31
	2000–2400	31	31	25	49	49	41	41	25
VI	0000–0400	49	49	31	49	31	31	25	25
	0400–0800	49	49	49	31	19	31	19	19
	0800–1200	49	49	25	19	16	19	25	31
	1200–1600	25	31	19	19	25	25	31	31
	1600–2000	25	25	19	41	31	41	41	49
	2000–2400	31	31	31	49	41	49	49	41
VII	0000–0400	25	19	19	41	31	19	25	19
	0400–0800	31	31	25	25	31	19	25	16
	0800–1200	49	49	31	25	19	25	25	25
	1200–1600	31	49	31	25	25	31	41	25
	1600–2000	31	31	25	41	41	41	49	41
	2000–2400	31	25	25	49	41	49	49	31
VIII	0000–0400	19	19	19	31	31	31	19	31
	0400–0800	25	25	25	31	25	25	19	25
	0800–1200	49	31	25	31	19	19	19	25
	1200–1600	31	31	31	19	19	19	25	41
	1600–2000	25	31	25	25	25	41	41	49
	2000–2400	25	25	25	31	31	41	41	49

* See legend on page 16.

Table 1-5. Favorable Shortwave Broadcast Bands (March–April 1979)

Listener's Area*	Time (UTC)	Broadcasting Station Location*							
		I	II	III	IV	V	VI	VII	VIII
I	0000–0400	49	31	31	49	31	31	25	19
	0400–0800	49	49	49	49	49	31	31	25
	0800–1200	49	49	49	31	49	49	31	31
	1200–1600	49	31	16	31	19	19	31	25
	1600–2000	25	25	16	19	16	16	25	19
	2000–2400	25	19	25	19	19	31	19	16
II	0000–0400	31	49	25	49	31	31	19	19
	0400–0800	49	49	49	49	31	31	19	19
	0800–1200	49	49	31	49	31	31	31	25
	1200–1600	31	49	31	31	19	19	25	31
	1600–2000	25	25	19	19	16	16	25	49
	2000–2400	19	25	16	25	16	16	19	19
III	0000–0400	31	25	49	41	31	31	25	19
	0400–0800	49	49	49	49	31	31	31	19
	0800–1200	49	31	49	31	31	25	31	25
	1200–1600	16	31	49	19	19	19	25	19
	1600–2000	16	19	25	16	16	19	19	19
	2000–2400	25	16	25	19	19	16	19	19
IV	0000–0400	49	49	41	49	31	41	31	31
	0400–0800	49	49	49	49	31	31	31	25
	0800–1200	31	49	31	31	19	19	19	19
	1200–1600	31	31	19	31	16	16	19	19
	1600–2000	19	19	16	31	19	19	25	25
	2000–2400	19	25	19	41	25	19	25	31
V	0000–0400	31	31	31	31	31	31	31	25
	0400–0800	49	31	31	31	25	19	19	19
	0800–1200	49	31	31	19	19	16	16	19
	1200–1600	19	19	19	16	25	19	19	31
	1600–2000	16	16	16	19	25	31	31	31
	2000–2400	19	16	19	25	25	31	31	25
VI	0000–0400	31	31	31	41	31	31	25	25
	0400–0800	31	31	31	31	19	25	19	19
	0800–1200	49	31	25	19	16	25	25	19
	1200–1600	19	19	19	16	16	31	31	31
	1600–2000	16	16	19	19	25	41	49	31
	2000–2400	31	16	16	19	31	49	41	25
VII	0000–0400	25	19	25	31	49	25	25	19
	0400–0800	31	19	31	31	31	19	19	16
	0800–1200	31	31	31	19	19	25	25	25
	1200–1600	31	25	25	19	19	31	41	31
	1600–2000	25	25	19	25	19	31	41	31
	2000–2400	19	19	19	25	31	31	41	31
VIII	0000–0400	19	19	19	31	25	31	19	31
	0400–0800	25	19	19	25	25	19	16	25
	0800–1200	31	25	25	19	19	19	19	25
	1200–1600	25	31	19	19	16	25	25	31
	1600–2000	19	49	19	25	19	31	31	31
	2000–2400	16	19	19	31	25	41	31	31

* See legend on page 16.

Table 1-6. Favorable Shortwave Broadcast Bands (May–August 1979)

Listener's Area*	Time (UTC)	Broadcasting Station Location*							
		I	II	III	IV	V	VI	VII	VIII
I	0000–0400	31	25	25	25	25	25	19	19
	0400–0800	49	31	31	25	31	19	25	25
	0800–1200	49	49	31	25	31	19	25	25
	1200–1600	31	31	25	25	19	16	19	25
	1600–2000	25	25	19	19	16	16	16	19
	2000–2400	19	19	19	19	19	19	19	19
II	0000–0400	25	25	25	25	25	25	19	16
	0400–0800	31	31	31	31	25	25	19	25
	0800–1200	49	49	49	31	25	25	19	31
	1200–1600	31	49	25	25	25	19	25	31
	1600–2000	25	31	19	19	19	19	25	19
	2000–2400	19	19	19	19	19	19	19	19
III	0000–0400	31	25	49	31	31	25	19	19
	0400–0800	31	25	49	31	31	25	19	19
	0800–1200	49	31	49	31	31	25	25	19
	1200–1600	25	31	31	19	19	19	19	31
	1600–2000	19	19	25	19	16	19	16	19
	2000–2400	19	19	31	25	19	19	16	19
IV	0000–0400	31	31	31	49	31	31	31	31
	0400–0800	31	31	31	41	31	31	25	25
	0800–1200	31	31	25	31	19	19	19	19
	1200–1600	25	25	19	31	19	16	19	19
	1600–2000	19	19	19	31	19	16	19	19
	2000–2400	25	19	25	31	25	25	31	31
V	0000–0400	31	31	31	31	49	31	31	19
	0400–0800	31	31	25	25	25	19	19	19
	0800–1200	25	31	19	19	25	19	16	25
	1200–1600	19	19	19	16	25	16	16	31
	1600–2000	19	19	16	19	31	31	25	31
	2000–2400	25	25	25	31	49	31	31	25
VI	0000–0400	25	25	25	31	31	31	25	25
	0400–0800	31	31	25	25	19	25	19	19
	0800–1200	25	25	25	16	19	25	19	25
	1200–1600	25	25	19	19	16	31	31	25
	1600–2000	19	19	19	25	31	41	25	25
	2000–2400	25	25	19	31	31	49	31	31
VII	0000–0400	19	19	19	31	31	25	25	16
	0400–0800	25	25	25	25	25	19	31	25
	0800–1200	25	31	25	25	16	19	25	25
	1200–1600	25	31	25	19	19	19	41	31
	1600–2000	25	25	25	31	25	25	49	31
	2000–2400	19	19	19	31	25	25	31	31
VIII	0000–0400	19	19	19	25	31	31	19	31
	0400–0800	25	25	19	31	25	25	19	31
	0800–1200	25	31	25	19	19	19	19	31
	1200–1600	31	31	25	19	19	25	31	31
	1600–2000	25	31	25	31	19	31	31	31
	2000–2400	19	25	25	31	25	31	31	49

* See legend on page 16.

Table 1-7. Favorable Shortwave Broadcast Bands (Sept–Oct 1979)

Listener's Area*	Time (UTC)	Broadcasting Station Location*							
		I	II	III	IV	V	VI	VII	VIII
I	0000–0400	49	25	31	19	31	31	25	19
	0400–0800	49	49	31	49	31	31	31	25
	0800–1200	49	49	49	49	31	31	31	25
	1200–1600	49	31	25	25	25	25	31	31
	1600–2000	25	25	19	19	19	19	31	19
	2000–2400	25	19	19	19	19	19	25	19
II	0000–0400	31	31	25	31	31	31	19	19
	0400–0800	49	49	31	49	31	25	19	19
	0800–1200	49	49	31	31	31	25	31	25
	1200–1600	31	49	31	31	25	25	31	31
	1600–2000	25	31	19	25	19	19	31	31
	2000–2400	19	19	16	31	16	25	19	19
III	0000–0400	49	25	25	41	31	31	19	19
	0400–0800	31	31	49	41	49	31	25	19
	0800–1200	31	31	49	31	31	25	31	19
	1200–1600	19	31	31	19	19	19	31	25
	1600–2000	16	19	25	16	16	19	25	25
	2000–2400	31	19	19	25	16	25	25	19
IV	0000–0400	31	31	49	49	31	31	31	31
	0400–0800	49	31	31	49	31	31	31	25
	0800–1200	31	49	31	31	19	19	19	19
	1200–1600	25	31	25	31	16	19	16	19
	1600–2000	19	25	19	31	19	25	31	25
	2000–2400	19	25	19	49	25	25	31	25
V	0000–0400	49	31	31	31	31	31	31	25
	0400–0800	31	31	25	25	25	19	19	19
	0800–1200	25	31	16	19	19	16	16	19
	1200–1600	19	25	16	16	25	16	16	31
	1600–2000	16	19	16	31	31	19	31	31
	2000–2400	31	25	19	31	31	31	31	25
VI	0000–0400	31	31	25	41	41	31	25	25
	0400–0800	31	31	25	31	19	25	19	19
	0800–1200	31	31	25	19	19	25	25	31
	1200–1600	25	25	19	19	16	31	31	31
	1600–2000	19	19	16	25	16	31	31	31
	2000–2400	25	25	25	41	41	31	31	25
VII	0000–0400	25	19	19	31	41	25	25	19
	0400–0800	31	25	25	25	31	19	19	16
	0800–1200	31	31	31	19	19	25	25	25
	1200–1600	31	31	31	19	19	31	41	31
	1600–2000	25	31	25	31	31	31	41	41
	2000–2400	25	25	25	31	31	31	41	31
VIII	0000–0400	19	16	19	31	25	31	19	31
	0400–0800	31	16	19	31	25	19	16	25
	0800–1200	31	25	25	25	19	19	16	25
	1200–1600	31	31	25	19	19	25	25	31
	1600–2000	25	31	25	19	19	31	31	31
	2000–2400	19	25	25	31	25	31	31	31

* See legend on page 16.

Table 1-8. Favorable Shortwave Broadcast Bands (Nov 1979–Feb 1980)

Listener's Area*	Time (UTC)	Broadcasting Station Location*							
		I	II	III	IV	V	VI	VII	VIII
I	0000–0400	31	31	31	49	31	31	25	25
	0400–0800	31	49	31	49	31	31	31	25
	0800–1200	31	49	31	31	31	31	31	31
	1200–1600	31	31	25	25	25	25	31	31
	1600–2000	19	25	16	19	19	19	31	25
	2000–2400	19	19	19	31	31	19	19	25
II	0000–0400	31	31	31	31	31	25	19	19
	0400–0800	31	31	31	49	31	25	31	31
	0800–1200	31	31	49	49	31	25	31	31
	1200–1600	31	31	31	31	25	25	49	31
	1600–2000	25	31	19	25	19	25	31	31
	2000–2400	19	25	19	25	25	31	25	16
III	0000–0400	49	31	31	49	31	49	19	19
	0400–0800	31	49	31	31	31	31	31	25
	0800–1200	31	31	31	31	31	31	31	25
	1200–1600	25	19	31	19	19	19	31	25
	1600–2000	19	19	25	16	16	19	25	19
	2000–2400	19	19	25	19	25	31	25	19
IV	0000–0400	31	49	49	49	41	41	41	41
	0400–0800	31	31	31	49	31	41	31	31
	0800–1200	31	31	31	49	19	19	19	31
	1200–1600	31	31	25	31	19	19	19	25
	1600–2000	19	31	19	31	19	31	41	25
	2000–2400	31	25	19	49	25	41	41	31
V	0000–0400	49	49	31	41	31	31	41	25
	0400–0800	31	31	31	25	25	19	25	19
	0800–1200	31	31	25	19	25	16	19	25
	1200–1600	19	25	16	16	25	16	19	25
	1600–2000	25	19	16	31	31	31	41	31
	2000–2400	25	19	25	31	31	31	41	25
VI	0000–0400	31	31	31	31	31	31	25	25
	0400–0800	31	31	31	31	19	31	19	19
	0800–1200	31	31	25	19	16	19	25	31
	1200–1600	25	31	19	19	16	25	25	31
	1600–2000	25	25	19	19	31	25	31	31
	2000–2400	31	25	31	31	31	31	31	41
VII	0000–0400	25	19	19	31	31	19	25	19
	0400–0800	31	31	25	25	31	19	25	16
	0800–1200	31	31	31	25	19	25	25	25
	1200–1600	31	31	31	25	25	25	41	25
	1600–2000	31	31	25	31	31	31	41	31
	2000–2400	25	25	25	31	31	31	41	25
VIII	0000–0400	19	19	19	31	31	31	19	31
	0400–0800	25	25	25	31	25	25	19	25
	0800–1200	31	31	25	25	25	19	19	25
	1200–1600	31	31	25	19	19	19	25	31
	1600–2000	25	31	31	19	19	19	31	31
	2000–2400	25	25	25	31	25	31	31	31

* See legend on page 16.

the average nontechnical shortwave listener can use, and, with a little practice, will be able to perform short-term ionospheric predictions with a high degree of accuracy.

As stated earlier, the ionosphere is formed by ultraviolet radiation from the sun. The stronger the radiation, the more dense the ionosphere will be; the weaker the radiation, the less dense the ionosphere will be. Thus the stronger the radiation layer, the better will be the shortwave propagation. Solar flux is a measure of the level of solar radiation and is thus an indication of the general state of the ionosphere. The use of solar flux as a measure of daily solar activity is now preferred rather than the daily sunspot count because solar flux has been found to be more direct and objective. It is also much more sensitive to change than the sunspot count. During the present low solar activity, daily solar flux levels will generally range between 70 and 100. The solar flux level is monitored at numerous observatories throughout the world.

While ultraviolet solar radiation produces the ionosphere, another type of solar radiation, called *solar particle radiation,* tends to weaken or deteriorate the ionosphere, or in some instances make it disappear completely. Because solar particle radiation also affects the earth's magnetic field, its level can be determined with the use of certain electronic instruments.

The earth's magnetic field is also monitored by several observatories throughout the world, and is reported as a worldwide planetary A index (A_p), which is the daily *average* measured by *all* stations. Since one full day is required to determine the A_p, it is not a real-time index.

Recently, the National Bureau of Standards radio station WWV began broadcasting both the A_p index (for the previous day) and a *real-time* three-hourly K index at eighteen minutes past each hour. The K and A indices are related approximately as follows:

$$K = 0 \quad 1 \quad 2 \quad 3 \quad 4 \quad 5 \quad 6 \quad 7 \quad 8 \quad 9$$
$$A_p = 0 \quad 4 \quad 7 \quad 15 \quad 27 \quad 48 \quad 80 \quad 140 \quad 240 \quad 400$$

The K index varies over a scale of 0 to 9. The higher the K value, the greater is the influx of solar particles, which in turn causes weaker signals and increased noise and fading conditions. Solar flux indicates the degree of ionization in the earth's atmosphere, and the A or K index measures the activity of the earth's magnetic field. Both taken together give a relatively accurate picture of overall ionospheric propagation conditions.

Fig. 1-1 can be used to determine the high-frequency ionospheric propagation conditions. For example, if the solar flux level is 80 and the geomagnetic index K is 2 (or an A_p index of 7), one can expect high "high normal" conditions.

With few exceptions, the *higher* the value of solar flux and the *lower* the value of magnetic activity, the better will be the ionospheric propagation. Conversely, the *lower* the solar flux and the *higher* the magnetic activity, the poorer will be the receiving conditions.

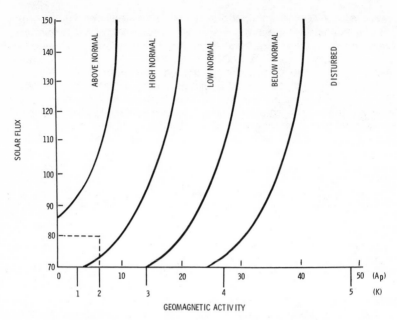

Fig. 1-1. Solar flux index versus geomagnetic activity.

A typical WWV radio broadcast is as follows: "Solar terrestrial indices for 11 June follow: solar flux index 85; *A* index 4. Repeat; solar flux index 85; *A* index 4. The *K* index at 1800 UTC 12 June was 1. Solar terrestrial conditions for the last 24 hours follow: solar activity was very low; the geomagnetic field was quiet. The forecast for the next 24 hours follows: solar activity will be low; the geomagnetic field will be unsettled."

The preceding typical WWV broadcast can be translated as follows: Since the broadcast was received on 12 June UTC, the solar flux reading of 85 and a A_p index of 4 was for the *previous* day. The current or real-time observation of the *K* index was taken at Boulder, Colorado on the current date at 1800 hours UTC. Using Fig. 1-1, both the previous day's reading and the present observation indicate a "high normal" condition, or good radio reception. However, a slight deterioration can be expected in the immediate future.

If the shortwave radio enthusiast maintains a daily log of WWV broadcasts and simultaneously makes observations of receiving conditions in his location, he will soon be able to make short-term ionospheric predictions with a high degree of accuracy. These observations are not only of great use to the swl'er, but can add to the enjoyment of his chosen hobby as well.

Due to the many variables involved and solar uncertainties, it is extremely difficult to predict shortwave propagation conditions beyond 18 months. For example, the next sunspot cycle is already one year late in starting. Since solar activity is responsible for the ionospheric changes, the reader can appreciate the complexities involved. Therefore, it is suggested that if the solar activity has increased appreciably during the early months of 1979 as indicated by the broadcasts from radio station WWV, all predicted wavelength bands in Tables 1-1 through 1-8 should be advanced by a factor of one; i.e., if the predicted band to use for best reception between areas for a specific time is the 31-meter band, the reader should also check the 25-meter band—reception might be better.

2

Stations by
Country and City

AFGHANISTAN

Kabul (35°00′N, 69°00′E)

Ann: "Da Radio Afghanistan Kabul Dai."
Int: A folk song played on a flute.

Stations:

4775 kHz, 100 kW
15195 kHz, 100 kW

Broadcasts:

Time (UTC)	Freq (kHz)
1130–1200	15195
1400–1430	4775

ALBANIA

Tirana (41°00′N, 20°00′E)

Ann: "This is Tirana calling."
Int: A few musical notes played on two trumpets.

Stations:

5945 kHz, 120 kW	9480 kHz, 120 kW
6200 kHz, 120 kW	9500 kHz, 500 kW
7065 kHz, 120 kW	9775 kHz, 100 kW
7070 kHz, 100 kW	11985 kHz, 100 kW
7300 kHz, 100 kW	

Broadcasts:

Time (UTC)	Freq (kHz)
0000–0030	(I) 7065, 9775
0130–0200	(I) 6200, 7070, 7300
0230–0300	(I) 6200, 7300
0330–0400	(I) 6200, 7300
0430–0500	(I) 5945, 7300
0630–0700	7065, 9500
0700–0730	(II) 9500, 11985
0930–1000	(II) 9500, 11985
1100–1130	(I) 9500, 11985
1400–1430	(III) 9500, 11985
1530–1600	(IV) 9480, 11985
1630–1700	7065, 9480
1730–1800	(IV) 7065, 9500
1830–1900	7065, 9480
1930–2000	(IV) 7065, 9500
2030–2100	7065, 9480
2200–2230	7065, 9480

Legend:

(I) Beamed to North America
(II) Beamed to Australia and New Zealand
(III) Beamed to Asia
(IV) Beamed to Africa

ALGERIA (Democratic and Popular Republic of)

Algiers: (36°42′N, 3°11′E)

Ann: "Ici Alger Radiodiffusion Television de la Republique Algerienne Democratique et Populaire."

Stations:

7245 kHz, 100 kW	11910 kHz, 100 kW
9610 kHz, 100 kW	15420 kHz, 100 kW

Broadcasts:

Time (UTC)	Freq (kHz)
1900–2000	7245, 9610, 11910, 15420

ANGOLA

Luanda (9°00′S, 12°00′E)

Ann: "This is the national radio station of Angola."
Int: A few bars of the national anthem.

Stations:

9535 kHz, 100 kW	11875 kHz, 100 kW

Broadcasts:

Time (UTC)	Freq (kHz)
1130–1200	(a) 9535
1435–1515	9535, 11875

Legend:

(a) Weekdays only

ARGENTINA

Buenos Aires (34°36′S, 58°22′W)

Ann: "This is R.A.E. calling."

Stations:

9690 kHz, 100 kW   11710 kHz, 100 kW

Broadcasts:

Time (UTC)	Freq (kHz)
0300–0400	(I) 9690
0600–0700	(II) 9690
2300–2400	(III) 11710

Legend:

(I) Beamed to North America (East)
(II) Beamed to North America (West)
(III) Beamed to Europe, North Africa, and Middle East

AUSTRALIA

Melbourne (38°00′S, 145°00′E)

Ann: "This is Radio Australia, transmitting the overseas service from the Melbourne studios of the Australian Broadcasting Commission. You are tuned to Radio Austrialia, the overseas service of the A.B.C."
Int: "Waltzing Matilda" played on a celesta.

Stations:

5995 kHz, 100 kW	9575 kHz, 100 kW
6005 kHz, 100 kW	9580 kHz, 100 kW
6035 kHz, 100 kW	9600 kHz, 100 kW
6045 kHz, 100 kW	9760 kHz, 100 kW
7240 kHz, 100 kW	11705 kHz, 100 kW
9515 kHz, 100 kW	11725 kHz, 100 kW
9520 kHz, 100 kW	11740 kHz, 100 kW
9540 kHz, 50 kW	11790 kHz, 100 kW
9550 kHz, 50 kW	11810 kHz, 100 kW
9570 kHz, 50 kW	11840 kHz, 250 kW

11935 kHz, 100 kW
15140 kHz, 20 kW
15180 kHz, 100 kW
15240 kHz, 50 kW
15290 kHz, 20 kW
15320 kHz, 100 kW

15355 kHz, 100 kW
15410 kHz, 50 kW
17795 kHz, 50 kW
17870 kHz, 100 kW
21570 kHz, 100 kW

Broadcasts:

Beamed to Pacific Islands

Time (UTC)	Freq (kHz)
0000–0600	15140
0000–0700	15180
0000–0730	15240
0000–0730	15320
0400–0930	11740
0630–0830	9515
0700–0900	9570
0700–2000	5995
0800–1730	9540
1800–2000	6045
1800–2030	9540
1800–2200	9580
1800–2200	11810
1900–2330	11725
2030–2200	11840
2100–2400	15240
2100–2400	15180
2300–2400	15140

Beamed to South and Southeast Asia

0000–0100	11705
0100–0930	17870
0130–1000	15410
0800–1000	21570
0800–1730	11705
1000–1730	9550
1500–1730	7240
1530–1730	6005
2100–2230	6035
2100–2400	9520
2230–2400	11705

Beamed to New Guinea Islands

0000–0200	11790
0130–0900	15355
0700–0800	9760
0700–0900	11790
1130–1730	7240
1800–2030	7240

2000–2200	9600
2100–2400	11790

Beamed to Asia and Northwest Pacific Area

0130–0900	15355
1100–1230	11740
1130–1300	9575
1400–1730	9575

Beamed to North America

0100–0300	15320, 17795
1100–1300	9580

Beamed to Europe

0700–0900	9570, 11740

Beamed to Africa

0400–0500	11935, 15290
0600–0630	11935, 15290

AUSTRIA

Vienna (48°00′N, 16°28′E)

Ann: "Osterreich auf Kurzwelle."
Int: First few bars of the "Blue Danube" waltz.

Stations:

6015 kHz, 100 kW	9770 kHz, 100 kW
6155 kHz, 100 kW	

Broadcasts:

Time (UTC)	Freq (kHz)
0330–0500	6155
1200–1600	9770
0430–0500	6015

BANGLADESH

Dacca (23°27′N, 90°12′E)

Ann: "This is the general overseas service of Radio Bangladesh."
Int: A portion of a folk song played on a violin and ta-npura.

Stations:

11890 kHz, 100 kW	15400 kHz, 100 kW
11940 kHz, 100 kW	17890 kHz, 100 kW
15180 kHz, 100 kW	21685 kHz, 100 kW
15270 kHz, 100 kW	

Broadcasts:

Time (UTC)	Freq (kHz)
0445–0510	15400, 17890, 21685
1230–1300	11890, 15270
1815–1915	11940, 15180

BELGIUM

Brussels (50°44′N, 4°34′E)

Ann: "This is Brussels, the Belgian Radio and Television Overseas Service."
Int: A few bars of the national anthem.

Stations:

6080 kHz, 100 kW	9745 kHz, 100 kW
9725 kHz, 100 kW	11940 kHz, 250 kW

Broadcasts

Time (UTC)	Freq (kHz)
0015–0045	(I) 6080, 9725
1715–1800	(II) 9745, 11940

Legend:

(I) Beamed to North America
(II) Beamed to Africa

BELIZE (Formerly British Honduras)

Belize (17°30′N, 88°30′W)

Ann: "This is Radio Belize, the voice of the new Central American nation of Belize in the heart of the Caribbean basin."

Station:

3300 kHz, 1 kW

Broadcasts:

Time (UTC)	Freq (kHz)
0030–0500	3300
1200–1600	3300
1700–2200	3300

BOTSWANA

Gaborone (24°42′S, 25°58′E)

Ann: "This is Radio Botswana, broadcasting from Gaborone."
Int: National anthem.

Stations:

3356 kHz, 10 kW	5965 kHz, 10 kW
4845 kHz, 10 kW	

Broadcasts:

Time (UTC)	Freq (kHz)
0400–0630	3356, 4845
0900–1300	5965
1500–2100	3356, 4845

BRUNEI

Bandar Seri Begawan (05°00′N, 115°00′E)

Ann: "This is Radio Television Brunei."
Int: Sound of a native drum.

Station:

7215 kHz, 10 kW

Broadcasts:

Time (UTC)	Freq (kHz)
0000–0030	7215
0300–0500	7215
1200–1430	7215
2300–2400	7215

BULGARIA

Sofia (42°00′N, 23°00′E)

Ann: "This is Sofia, Bulgaria calling."
Int: A march played on an organ.

Stations:

6070 kHz, 100 kW	9705 kHz, 50 kW
7115 kHz, 100 kW	11765 kHz, 50 kW
9560 kHz, 100 kW	17825 kHz, 50 kW
9700 kHz, 120 kW	

Broadcasts:

Time (UTC)	Freq (kHz)
0000–0100	(I) 9705
0430–0500	(I) 7115
1900–1930	(II) 11765, 17825
1930–2000	(III) 6070, 9700
2100–2200	(II) 9560, 11765, 17825
2130–2200	(III) 9700

Legend:

  (I) Beamed to North America
  (II) Beamed to Africa
 (III) Beamed to Europe

BURMA

Rangoon (16°52′N, 96°10′E)

Ann: "This is the Burma Broadcasting Service."
Int: Orchestral music.

Stations:

5040 kHz, 50 kW	9725 kHz, 50 kW
7185 kHz, 50 kW	

Broadcasts:

Time (UTC)	Freq (kHz)
0200–0230	7185
0700–0730	9725
1430–1600	5040

BURUNDI

Bujumbura (3°23′S, 29°21′E)

Ann: "This is Radio Cordac, Bujumbura, Burundi, the Central African Broadcasting Co., Inc."
Int: Religious song played on a guitar.

Stations:

  3973 kHz, 2.5 kW
  4900 kHz, 2.5 kW

Broadcasts:

Time (UTC)	Freq (kHz)
0500–0515	(a) 3973, 4900
0700–0730	(b) 3973, 4900
1630–1700	(b) 3973, 4900

Legend:

 (a) Weekdays only
 (b) Sundays only

CAMEROON

Buea (4°09′N, 0°14′E)

Ann: "This is Buea, the provincial station of Radio Cameroon."
Int: Music played on a balaphon.

Stations:

3970 kHz, 8 kW 6005 kHz, 4 kW

Broadcasts:

Time (UTC)	Freq (kHz)
0430–0800	(a) 3970
1100–1630	(a) 6005
1630–2200	(a) 3970

Legend:

(a) French and English language

Garoua (9°18′N, 13°25′E)

Ann: "Ici Garoua Radiodiffusion de la Republique Unie du Cameroun, Station Provinciale."

Station:

5010 kHz, 30 kW

Broadcasts:

Time (UTC)	Freq (kHz)
1830–1845	5010

CANADA

Montreal (45°52′N, 64°19′W)

Ann: "This is Radio Canada International."
Int: Four musical notes played on an electric organ.

Stations:

5970 kHz, 250 kW	9625 kHz, 250 kW
6040 kHz, 250 kW	9650 kHz, 250 kW
6085 kHz, 250 kW	9655 kHz, 250 kW
6125 kHz, 250 kW	11720 kHz, 250 kW
6135 kHz, 250 kW	11825 kHz, 250 kW
6140 kHz, 250 kW	11855 kHz, 250 kW
6195 kHz, 250 kW	11940 kHz, 250 kW
7155 kHz, 250 kW	15325 kHz, 250 kW
9560 kHz, 250 kW	17820 kHz, 250 kW

Broadcasts:

Beamed to Western Europe

Time (UTC)	Freq (kHz)
0620–0640	6125, 6140, 7155, 9655, 11720
0700–0720	6125, 6140, 7155, 9655, 11720
0740–0800	6125, 6140, 7155, 9655, 11720
1145–1215	(a) 9560, 11720
1400–1500	(b) 6195

| 2100–2200 | 9625, 11855, 15325 |
| 2200–2230 | 11855, 15325 |

Beamed to USA

0100–0200	6085, 9650
0400–0600	6135, 9655
1115–1215	5970, 9655
2200–2400	(c) 6040

Beamed to Caribbean Area

| 0100–0157 | 9650, 11940 |
| 1115–1215 | 9655, 11825 |

Beamed to Africa

0620–0640	11720
0700–0720	11720
0740–0800	11720
1800–1900	11855, 15325, 17820

Beamed to South Pacific Area

| 1000–1100 | 5970, 9625 |

Legend:

(a) Armed Forces Service news and sports
(b) Sunday only
(c) Monday through Friday only

CENTRAL AFRICAN REPUBLIC

Bangui (4°21′N, 18°35′E)

Ann: "Ici la Radiodiffusion Nationale Centrafricaine qui emet de Bangui."
Int: National Anthem.

Station:

5035 kHz, 100 kW

Broadcasts:

| *Time (UTC)* | *Freq (kHz)* |
| 2030–2045 | 5035 |

CHAD

Ndjamena (Fort-Lamy) (12°08′N, 15°03′E)

Ann: "Ici Ndjamena Radiodiffusion Nationale Tchadienne emettant dans les bandes."
Int: Music on balaphon.

Stations:

4904 kHz, 30 kW   9615 kHz, 4 kW
7120 kHz, 30 kW

Broadcasts:

Time (UTC)	Freq (kHz)
0430–0600	(a) 4904
0430–0800	(a) (b) 4904
1230–1610	(a) 7120, 9615
0800–1610	(a) (b) 7120, 9615
1610–2130	(a) 4904
1610–2300	(a) (c) 4904

Legend:

(a) French language only
(b) Sundays only
(c) Saturdays only

CHILE

Santiago (33°00'S, 71°00'W)

Ann: "Radio Nacional, La Voz de Chile"

Station:

9565 kHz, 10 kW

Broadcasts:

Time (UTC)	Freq (kHz)
0100–0130	9565
0210–0230	9565
0310–0350	9565
1050–1110	9565
1210–1230	9565
1330–1350	9565
2250–2310	9565

CHINA (People's Republic of; Mainland)

Peking (40°00'N, 116°30'E)

Ann: "This is Radio Peking."
Int: The first few bars of "The East is Red"; sign off with "The Internationale."

Stations:

6290 kHz, 120 kW   7155 kHz, 120 kW
6860 kHz, 120 kW   7315 kHz, 120 kW
7035 kHz, 120 kW   7590 kHz, 120 kW
7120 kHz, 120 kW   7620 kHz, 120 kW

9030 kHz, 120 kW	11650 kHz, 120 kW
9380 kHz, 120 kW	11675 kHz, 120 kW
9460 kHz, 120 kW	11685 kHz, 120 kW
9470 kHz, 120 kW	11945 kHz, 120 kW
9780 kHz, 120 kW	12010 kHz, 120 kW
9860 kHz, 120 kW	12055 kHz, 120 kW
9940 kHz, 120 kW	15060 kHz, 120 kW
11445 kHz, 120 kW	15270 kHz, 120 kW

Broadcasts:

Beamed to North America (East)

Time (UTC)	Freq (kHz)
0000–0100	11675, 11945, 15060
0100–0200	7120, 9780, 9940, 11945, 12055
0200–0300	9940, 12010, 12055
0300–0400	7120, 9780
1200–1300	11685

Beamed to North America (West)

0300–0500	9460, 11650, 12055, 15060

Beamed to Australia and New Zealand

0830–1030	7035, 9380, 9460, 11445, 15060

Beamed to Southeast Asia

1200–1400	6290, 7155, 11650, 15270

Beamed to South Asia

1400–1600	7315, 9860, 11650

Beamed to East and South Africa

1600–1800	7620, 9860

Beamed to West and North Africa

1930–2130	7620, 9470, 11650

Beamed to Europe

2030–2230	6860, 7590, 9030

COMORO ISLANDS

Moroni (11°42′S, 43°15′E)

Ann: "Comores-Internationale ORTF."
Int: Folk song.

Stations:

3331 kHz, 4 kW	7260 kHz, 4 kW

Broadcasts:

Time (UTC)	Freq (kHz)
0400–0430	(a) 3331, 7260
0900–1030	(a) 3331, 7260
0500–1030	(a) (b) 3331, 7260
1500–1730	(a) 3331, 7260

Legend:

(a) French language
(b) Sunday only

CONGO (People's Republic of the)

Brazzaville (4°15′S, 15°18′E)

Ann: "Ici la voix de la Revolution Congolaise, Station Nationale de Radiodiffusion Television."
Int: Musical notes on a native instrument.

Stations:

4795 kHz, 4 kW	15190 kHz, 50 kW
9715 kHz, 50 kW	

Broadcasts:

Time (UTC)	Freq (kHz)
2115–2145	(a) (b) 4795, 9715, 15190
2150–2200	(a) (c) 4795, 9715, 15190
2100–2130	(a) (d) 4795, 9715, 15190
2235–2305	(a) (e) 4795, 9715, 15190

Legend:

(a) French and English language
(b) Monday through Thursday
(c) Friday only
(d) Saturday only
(e) Sunday only

CUBA

Havana (23°00′N, 82°30′W)

Ann: "This is Radio Havana, Cuba."
Int: National anthem.

Stations:

9525 kHz, 50 kW	11930 kHz, 100 kW
9685 kHz, 50 kW	15300 kHz, 50 kW
11725 kHz, 100 kW	17885 kHz, 50 kW
11760 kHz, 100 kW	

Broadcasts:

Time (UTC)	Freq (kHz)
0100–0450	(I) 9685
0100–0600	(I) 11725
0330–0600	(I) 11760
0630–0800	(I) 9525
2010–2140	(II) 17885
2050–2140	(I) 11930, 15300

Legend:

(I) Beamed to North, Central, and South America
(II) Beamed to Europe

CZECHOSLOVAKIA

Prague: (50°09′N, 15°09′E)

Ann: "This is Prague, Czechoslovakia."
Int: A few bars of the song "Forward Left."

Stations:

5930 kHz, 120 kW	9740 kHz, 200 kW
6055 kHz, 120 kW	11855 kHz, 200 kW
7245 kHz, 120 kW	11990 kHz, 200 kW
7345 kHz, 200 kW	15110 kHz, 120 kW
9505 kHz, 200 kW	15395 kHz, 120 kW
9540 kHz, 200 kW	17840 kHz, 120 kW
9605 kHz, 200 kW	21700 kHz, 120 kW
9630 kHz, 200 kW	

Broadcasts:

Beamed to North America

Time (UTC)	Freq (kHz)
0100–0200	9540, 9630, 9740
0300–0400	5930, 7345

Beamed to Europe

1630–1700	5930, 7345
1900–1930	5930, 7245, 7345
2000–2030	5930, 7345
2130–2200	6055

Beamed to Asia and Pacific Area

0730–0800	6055, 9505, 11855, 15395, 21700
0830–0930	6055, 9505
1430–1500	5930, 7345, 9605, 11990, 15110, 17840

Beamed to Africa

1730–1830	5930, 7345, 9605, 11990, 17840

DAHOMEY

Cotonou (6°21′N, 2°35′E)

Ann: "Ici Cotonou Radiodiffusion du Dahomey."
Int: Dahomey tam-tam.

Stations:

3270 kHz, 4 kW	7190 kHz, 30 kW
4870 kHz, 30 kW	

Broadcasts:

Time (UTC)	Freq (kHz)
0510–0700	(a) (b) 3270, 4870, 7190
1130–1300	(a) (b) 3270, 4870
1630–2200	(a) (b) 3270, 4870
0515–0730	(a) (c) 3270, 4870, 7190
1200–2300	(a) (c) 3270, 4870, 7190
0700–2300	(a) (d) 3270, 4870

Legend:

(a) French language
(b) Monday through Friday only
(c) Saturday only
(d) Sunday only

DJIBOUTI (People's Republic of)

Djibouti (11°30′N, 43°00′E)

Ann: "Ici Djibouti Office de Radiodiffusion Television Francaise."
Int: A few bars of a Somalian song.

Station:

4780 kHz, 4 kW

Broadcasts:

Time (UTC)	Freq (kHz)
0300–0600	(a) 4780
0500–2000	(a) (b) 4780
0900–2000	(a) 4780

Legend:

(a) Arabic language only
(b) Sundays only

ECUADOR

Quito (0°14′S, 78°20′W)

Ann: "This is the voice of the Andes, Quito Ecuador."
Int: The first few bars of a folk song.

Stations:

6095 kHz, 100 kW	15115 kHz, 50 kW
9560 kHz, 100 kW	15310 kHz, 50 kW
11745 kHz, 100 kW	17780 kHz, 50 kW
11915 kHz, 100 kW	

Broadcasts:

Time (UTC)	Freq (kHz)
0040–0700	(I) 6095, 9560, 11915
1230–1630	(II) 11745, 15115
1630–1800	(III) 15310, 17780
1900–2030	(IV) 15310

Legend:

(I) Beamed to North America
(II) Beamed to North and South America
(III) Beamed to Middle East
(IV) Beamed to Europe and Middle East

EGYPT

Cairo (30°16′N, 31°22′E)

Ann: "The Voice of the Arabs."

Stations:

9475 kHz, 100 kW	17920 kHz, 100 kW
9805 kHz, 100 kW	

Broadcasts:

Time (UTC)	Freq (kHz)
0200–0330	(I) 9475
1315–1430	(II) 17920
2145–2300	(III) 9805

Legend:

(I) Beamed to North America
(II) Beamed to Asia
(III) Beamed to Europe

EL SALVADOR

San Salvador (13°00′N, 89°00′W)

Stations:

5980 kHz, 5 kW	9555 kHz, 5 kW

Broadcasts:

Time (UTC)	Freq (kHz)
0000–0600	(a) 5980, 9555
1200–2400	(a) 5980, 9555

Legend:

(a) Spanish language only

EQUATORIAL GUINEA

Malabo (3°30'N, 9°00'E)

Ann: "Radio Malabo, emisora en la capital de la Republica de Guinea Ecuatorial."
Int: National Anthem

Station:

6250 kHz, 10 kW

Broadcasts:

Time (UTC)	Freq (kHz)
2030–2100	6250

ETHIOPIA

Addis Ababa (8°58'N, 38°43'E)

Ann: "This is the Voice of Revolutionary Ethiopia."
Int: Folk song played on a native flute.

Stations:

6185 kHz, 100 kW	8072 kHz, 100 kW
7165 kHz, 100 kW	9610 kHz, 100 kW
7180 kHz, 100 kW	16085 kHz, 100 kW

Broadcasts:

Time (UTC)	Freq (kHz)
1000–1030	6185, 7165, 8072, 9610
1500–1600	7165, 7180, 9610, 16085
1700–1800	6185, 7165, 9610

FINLAND

Helsinki (61°28'N, 21°52'E)

Ann: "This is the Finnish Broadcasting Company."
Int: Finnish folk song.

Stations:

6105 kHz, 250 kW	11755 kHz, 250 kW
6120 kHz, 100 kW	15110 kHz, 100 kW
9550 kHz, 100 kW	15265 kHz, 250 kW

Broadcasts:

Time (UTC)	*Freq (kHz)*
0300–0400	(I) (a) 6105
0930–1100	(II) 9550
1230–1400	(I) (a) 11755, 15110
1430–1630	(I) 11755, 15110
1700–1930	(II) 15265
2000–2130	(III) (a) 6120, 9550, 11755

Legend:

(I) Beamed to North America
(II) Beamed to Europe
(III) Beamed to Africa
(a) English and Finnish language

FRANCE

Paris (47°00′N, 2°00′E)

Ann: "Ici Paris, Radio France International, Cooperation Radio-phonique."
Int: French song, "Nous n'irons plus au bois."

Stations:

7165 kHz, 100 kW	11705 kHz, 100 kW
7285 kHz, 100 kW	11890 kHz, 500 kW
9505 kHz, 100 kW	11930 kHz, 500 kW
9585 kHz, 500 kW	15425 kHz, 500 kW
9695 kHz, 500 kW	

Broadcasts:

Time (UTC)	*Freq (kHz)*
1700–1800	(I) 7165, 7285, 9505, 9585, 9695, 11705, 11890, 11930, 15425

Legend:

(I) Beamed to Africa

GABON

Libreville (0°25′N, 9°28′E)

Ann: "Ici Libreville, Radiodiffusion Television Gabonaise la voix de la Renovation chaine Nationale."

Stations:

3330 kHz, 100 kW	7270 kHz, 20 kW
4777 kHz, 20 kW	

Broadcasts:

Time (UTC)	Freq (kHz)
0430–0630	(a) 3330, 4777
0630–1630	(a) 3330, 7270
1630–2300	(a) 3330, 4777

Legend:

(a) French language only

GERMANY (Democratic Republic; East)

Berlin (52°18'N, 13°37'E)

Ann: "This is Radio Berlin International, the Voice of the German Democratic Republic."
Int: National Anthem.

Stations:

6080 kHz, 100 kW	11890 kHz, 100 kW
6115 kHz, 100 kW	15125 kHz, 100 kW
7185 kHz, 100 kW	15145 kHz, 100 kW
7260 kHz, 100 kW	15170 kHz, 100 kW
7300 kHz, 100 kW	15250 kHz, 100 kW
9500 kHz, 100 kW	15320 kHz, 100 kW
9650 kHz, 100 kW	17700 kHz, 100 kW
9665 kHz, 100 kW	17800 kHz, 100 kW
9730 kHz, 100 kW	21465 kHz, 100 kW
11720 kHz, 100 kW	21540 kHz, 100 kW
11840 kHz, 100 kW	

Broadcasts:

Beamed to USA (East Coast)

Time (UTC)	Freq (kHz)
0100–0145	9730
0230–0315	9730

Beamed to USA (West Coast)

0330–0415	9650, 11840, 11890

Beamed to Europe

1930–2000	6080, 6115, 7185, 7300, 9730
2215–2245	7260

Beamed to Southeast Asia

0645–0730	17700, 21465
1200–1245	15125, 15320, 21540
1400–1445	15125, 21540
1530–1615	15125

Beamed to West Africa

0445–0530	9500
1800–1845	15250
2000–2045	9665, 15250

Beamed to East Africa

0445–0530	11720
1800–1845	15145

Beamed to Central Africa

1315–1400	17800

GERMANY (Federal Republic; West)

Cologne (50°57′N, 6°22′E)

Ann: "This is Radio Deutsche Welle, the voice of Germany."
Int: Two bars from "Fidelio," played on a celesta.

Stations:

5960 kHz, 100 kW	9735 kHz, 100 kW
6010 kHz, 100 kW	9760 kHz, 100 kW
6040 kHz, 100 kW	9765 kHz, 100 kW
6065 kHz, 100 kW	11765 kHz, 100 kW
6075 kHz, 100 kW	11785 kHz, 100 kW
6100 kHz, 100 kW	11810 kHz, 100 kW
6185 kHz, 100 kW	11850 kHz, 100 kW
7130 kHz, 100 kW	11865 kHz, 100 kW
7150 kHz, 100 kW	11905 kHz, 100 kW
7160 kHz, 100 kW	11925 kHz, 100 kW
7210 kHz, 100 kW	11970 kHz, 100 kW
7225 kHz, 100 kW	15275 kHz, 100 kW
7285 kHz, 100 kW	15410 kHz, 100 kW
9545 kHz, 100 kW	17715 kHz, 100 kW
9565 kHz, 100 kW	17765 kHz, 100 kW
9590 kHz, 100 kW	17780 kHz, 100 kW
9605 kHz, 100 kW	17825 kHz, 100 kW
9615 kHz, 100 kW	17875 kHz, 100 kW
9650 kHz, 100 kW	21500 kHz, 100 kW
9690 kHz, 100 kW	21540 kHz, 100 kW
9700 kHz, 100 kW	21600 kHz, 100 kW

Broadcasts:

Time (UTC)	Freq (kHz)
0130–0150	(I) 6010, 6040, 6075, 6100, 9565, 9590, 9605, 11865
0120–0220	(II) 6065, 7210, 7285, 9690
0430–0515	(III) 6065, 7150, 7225, 9565, 9765
0530–0550	(IV) 5960, 6100, 6185, 9545, 9760
0600–0630	(V) 9615, 9700, 11765, 11905, 15275
0930–1030	(VI) 9650, 11850, 15275, 17715, 17780, 17825, 21540
1045–1115	(III) 11785, 15410, 17765, 17875, 21500, 21600
1200–1245	(V) 15410, 17765, 17875, 21600
1300–1320	(I) 9605, 11810, 11970
1715–1745	(III) 7285, 9735, 11810, 11850
1720–1750	(II) 7160, 9590, 11785, 11925
1930–2000	(V) 9765, 11905
2100–2200	(VI) 7130, 9765

Legend:

- (I) Beamed to North America (East)
- (II) Beamed to South Asia
- (III) Beamed to Africa
- (IV) Beamed to North America (West)
- (V) Beamed to West Africa
- (VI) Beamed to East Asia, Australia, and New Zealand

GHANA

Accra (5°31′N, 0°10′E)

Ann: "This is the external service of Radio Ghana."
Int: First bars of the national anthem played on a guitar.

Stations:

6130 kHz, 100 kW	15285 kHz, 100 kW
9545 kHz, 100 kW	21545 kHz, 100 kW
11850 kHz, 250 kW	21720 kHz, 250 kW

Broadcasts:

Time (UTC)	Freq (kHz)
0700–0900	(I) 6130
1400–1430	(II) 21720
1445–1530	(II) (III) 15285; (IV) 21545, 21720
1600–1700	(I) 6130
1645–1730	(II) 15285
1815–1900	(II) 15285
2000–2100	(V) 11850
2000–2300	(I) 6130
2030–2200	(VI) 9545, 15285

Legend:

- (I) Beamed to West Africa
- (II) Beamed to East Africa
- (III) Beamed to South Africa
- (IV) Beamed to Australia and New Zealand
- (V) Beamed to North America and Caribbean
- (VI) Beamed to Europe

GREAT BRITAIN

London (52°19′N, 00°00′)

Ann: "This is the World Service of the BBC."

Int: Musical selection "Oranges and Lemons"; time signal every hour, Big Ben every quarter hour.

Stations:

Some stations listed below are located in London; others are located in various countries throughout the world, relaying BBC transmissions.

3915 kHz, 250 kW	9590 kHz, 250 kW
3952 kHz, 250 kW	9600 kHz, 250 kW
3989 kHz, 250 kW	9640 kHz, 250 kW
5070 kHz, 250 kW	9670 kHz, 250 kW
5960 kHz, 250 kW	9740 kHz, 250 kW
5975 kHz, 250 kW	9750 kHz, 250 kW
5999 kHz, 250 kW	9760 kHz, 250 kW
6005 kHz, 250 kW	9915 kHz, 250 kW
6050 kHz, 250 kW	11750 kHz, 250 kW
6140 kHz, 250 kW	11760 kHz, 250 kW
6175 kHz, 250 kW	11770 kHz, 250 kW
6180 kHz, 250 kW	11820 kHz, 250 kW
6195 kHz, 250 kW	11860 kHz, 250 kW
7105 kHz, 250 kW	11955 kHz, 250 kW
7120 kHz, 250 kW	12040 kHz, 250 kW
7135 kHz, 250 kW	12095 kHz, 250 kW
7140 kHz, 250 kW	15070 kHz, 250 kW
7150 kHz, 250 kW	15105 kHz, 250 kW
7180 kHz, 250 kW	15260 kHz, 250 kW
7185 kHz, 250 kW	15280 kHz, 250 kW
7210 kHz, 250 kW	15310 kHz, 250 kW
7230 kHz, 250 kW	15365 kHz, 250 kW
7270 kHz, 250 kW	15410 kHz, 250 kW
7325 kHz, 250 kW	15420 kHz, 250 kW
9410 kHz, 250 kW	15435 kHz, 250 kW
9510 kHz, 250 kW	17695 kHz, 250 kW
9540 kHz, 250 kW	17705 kHz, 250 kW
9570 kHz, 250 kW	17790 kHz, 250 kW
9580 kHz, 250 kW	17840 kHz, 250 kW

17880 kHz, 250 kW   21610 kHz, 250 kW
17885 kHz, 250 kW   21660 kHz, 250 kW
21470 kHz, 250 kW   21710 kHz, 250 kW

Broadcasts:

Beamed to North and Central America and Caribbean

Time (UTC)	Freq (kHz)
0000–0230	11750
1100–1330	6195
2000–2130	6195
1100–1330	15070
1100–1330	5999
1500–1630	17840
1500–1800	9580
1500–1800	15365
2000–2245	5960
2000–2330	11750
2000–2330	15260
2100–0330	9580
2300–0730	5975
2300–0730	6175
2300–0330	7325
2300–0330	9510

Beamed to South America

0030–0300	9510
2000–2230	5960
2000–0230	11750
2000–2330	15260
2115–0030	9580
2230–0330	6005
2245–0330	5975
2300–0330	9915

Beamed to Western Europe

0400–0730	6050
0400–0730	6180
0400–0730	7185
0500–2330	5975
0500–0730	15420
0530–0730	3952
0600–0730	6195
0600–0730	7230
0600–0730	9750
0900–1630	9750
0900–1830	12095
1600–2330	7120
1600–2330	9410

1700–2330	6180
1700–1800	6195
2000–2100	6195

Beamed to Northern Europe

0400–0630	6180
0400–0730	6050
0400–0730	15420
0530–0730	9410
0600–0730	7120
0600–0730	9750
0900–1630	12095
0900–1830	15070
0900–1630	9750
0900–2130	9410
1700–2130	6180
1730–2130	7120
1800–2130	5975

Beamed to Southwestern Europe

0500–0730	5975
0500–0730	7185
0500–0730	9580
0600–0730	12095
0900–1830	12095
0900–1530	9760
0900–2030	15070
1500–0030	9410
1700–0030	5975
1800–2330	7185
1830–2130	9750

Beamed to Central and Southeast Europe

0300–0730	6050
0300–0730	7185
0300–0730	9410
0400–0730	6180
0400–0730	11750
0400–0730	15420
0500–0730	15070
0600–0730	9750
0900–1330	17790
0900–1530	21710
0900–1630	9750
0900–2030	15070
0900–2330	9410
0900–2130	12095
1600–2330	7120
1700–2130	3989

1700–2330	6180
1700–0030	5975

Beamed to North and Northwest Africa

0400–0730	5975
0400–0730	7185
0400–0730	9410
0400–0730	11750
0500–0730	9580
0600–0730	12095
0700–2130	15070
0900–1830	12095
0900–1630	17705
0900–1330	21470
1030–1630	21710
1600–0030	9410
1700–0030	5975
1700–2330	7120
1700–2130	11750
1800–2400	7185
2230–0030	7130

Beamed to West and Central Africa

0400–0730	6005
0400–0730	7185
0400–0730	9580
0400–0730	11750
0700–0930	15070
0730–1630	15400
0730–0930	21710
0800–1130	11860
0900–1730	17705
1030–1630	21710
1300–2300	15070
1600–2230	9410
1700–2130	11750
1800–2300	6005

Beamed to East Africa

0300–0430	5975
0300–0630	7185
0300–0630	9410
0300–0630	9580
0400–0700	11750
0500–0730	15420
0600–0730	17885
0900–1300	21660
0900–1330	21470
0900–1630	17885

0900–1700	15420
1300–2030	15070
1600–2130	7120
1600–2130	9410
1600–1830	15105
1600–2030	15400
1700–2130	11750
1830–2030	11820

Beamed to South Africa

0400–0730	6005
0400–0600	7185
0400–0630	11750
0400–0730	15420
0430–0630	9580
0600–0730	17885
0700–0930	15070
0730–1800	15400
0900–1630	17885
1030–1630	21710
1430–1830	15070
1600–2300	9410
1600–1830	15105
1700–2130	11750
1700–2300	7120
1700–2030	6140
1830–2030	11820

Beamed to Middle East

0300–0430	5975
0300–0530	6050
0300–0430	11955
0300–0630	7185
0300–0730	9410
0400–0730	11750
0500–2030	15070
0600–1330	11760
0700–1530	21710
0900–1230	7140
0900–1330	17790
0900–1230	15420
1300–1630	15310
1300–2130	12095
1600–2330	7120
1600–2330	9410
1700–1830	3989
1700–2130	5975
2000–2130	3989

Beamed to U.S.S.R.

0300–0400	3952
0300–0630	6050
0300–0730	9410
0400–0630	6180
0500–0730	15420
0530–0730	7120
0900–1830	12095
0900–1330	17790
0900–1330	21610
0900–1830	15070
1300–2130	9410
1700–2130	6180
1730–2130	7120
1730–2130	5975

Beamed to South Atlantic

2230–0330	5975, 7130, 9915

Beamed to South Asia and Iran

0000–0330	7140
0200–0430	11955
0900–1230	15420
0900–1130	9670
0900–1630	15310
0900–1630	11750
0900–1830	6195
0900–1830	9740
1600–1830	11955
1600–2030	9410
1630–1830	7180
2300–0330	6195
2300–0030	7120
2300–0330	9410

Beamed to Indonesia

0900–1130	9670
0900–1830	6195
0900–1630	9740
0900–1630	11750
1500–1830	3915
2200–0030	6195
2200–0030	7120
2200–0030	9570

Beamed to Australia and New Zealand

0530–0930	11955
0600–0730	5975
0600–0930	7150

0600–0930	9640
0900–1630	9740
0900–1630	11750
0900–1330	15070
0900–1330	15310
0900–1130	17695
1100–1630	6195
1100–1630	12095
1430–1630	9410
2000–2130	5975
2000–2130	7120
2000–2230	9410
2000–2300	11750

Beamed to Southeast Asia

0000–0100	17880
0900–1000	17880
0900–1130	9670
0900–1130	15280
0900–1630	9740
0900–1630	11750
0900–1830	6195
1500–1830	3915
2200–0030	6195
2200–0030	7120
2200–0030	11955
2230–2330	9580
2300–0030	15435

Beamed to Asia (General Coverage)

0900–1830	15070
0900–1330	17790
0900–1330	21610
1300–1830	12095
1430–1830	9410
2200–0030	5975
2200–0030	9410

GUIANA (French)

Cayenne (5°00′N, 52°00′W)

Ann: "Ici Cayenne, Office de Radiodiffusion Television Francaise."
Int: Folk song played on a guitar.

Stations:

3385 kHz, 4 kW	6170 kHz, 4 kW

Broadcasts:

Time (UTC)	Freq (kHz)
0900–1200	(a) 3385
1200–2000	(a) 6170
2000–0145	(a) 3385

Legend:

(a) French language only

GUINEA (Republic of)

Conakry (9°32′N, 13°40′W)

Ann: "Ici la voix de la Revolution."
Int: A few bars of the national anthem played on a guitar.

Stations:

4910 kHz, 18 kW	9650 kHz, 100 kW
6155 kHz, 18 kW	11965 kHz, 100 kW
7125 kHz, 18 kW	

Broadcasts:

Time (UTC)	Freq (kHz)
0000–0830	(a) 4910, 6155, 7125, 9650, 11965
1230–2400	(a) 4910, 6155, 9650, 11965
1600–2400	(a) 7125

Legend:

(a) French language only

GUYANA (Republic)

Georgetown (6°49′N, 58°05′W)

Ann: "This is Action Radio; Guyana Broadcasting Service."

Stations:

3290 kHz, 10 kW	5950 kHz, 10 kW

Broadcasts:

Time (UTC)	Freq (kHz)
0730–1200	3290
1200–2215	5950
2215–0300	3290

Georgetown (6°49′N, 58°05′W)

Ann: "This is Radio Demerara, the voice of Guyana."

Stations:

3265 kHz, 2 kW	5980 kHz, 2 kW

Broadcasts:

Time (UTC)	Freq (kHz)
0815–0200	3265, 5980

HAITI

Cap Haitien (19°00′N, 73°00′W)

Ann: "This is Radio Station 4VEH, Cap Haitien."

Stations:

6120 kHz, 2 kW   11835 kHz, 2.5 kW
9770 kHz, 0.35 kW

Broadcasts:

Time (UTC)	Freq (kHz)
0000–0100	6120, 9770, 11835
1100–1400	6120, 9770, 11835

HONDURAS (Republic of)

Tegucigalpa (14°00′N, 87°00′W)

Ann: "This is the Baptist Home Mission Society."
Int: Hymn.

Station:

4820 kHz, 5 kW

Broadcasts:

Time (UTC)	Freq (kHz)
0300–0600	4820

HUNGARY

Budapest (47°00′N, 19°00′E)

Ann: "This is Budapest, Hungary."
Int: A few bars of the musical selection "1848."

Stations:

5965 kHz, 100 kW	15225 kHz, 100 kW
6000 kHz, 100 kW	15415 kHz, 100 kW
7155 kHz, 100 kW	17710 kHz, 100 kW
7200 kHz, 100 kW	17715 kHz, 100 kW
9585 kHz, 100 kW	17720 kHz, 100 kW
9655 kHz, 100 kW	17780 kHz, 100 kW
11910 kHz, 100 kW	17785 kHz, 100 kW
15160 kHz, 100 kW	21525 kHz, 100 kW

Broadcasts:

Time (UTC)	Freq (kHz)
0200–0230	(I) (a) 6000, 9585, 11910, 15225, 17710
0300–0330	(I) 6000, 9585, 11910, 15225, 17720
0400–0415	(I) (b) 6000, 9585, 11910, 15225, 17720
1030–1100	(II) 7155, 9585, 11910, 15160, 17715, 21525
1200–1240	(III) (c) 7155, 9585, 11910, 15160, 17785, 21525
1430–1500	(IV) (c) 7155, 9585, 11910, 15160, 17785, 21525
2130–2200	(III) 5965, 7200, 9655, 11910, 15415, 17780

Legend:

(I) Beamed to North America
(II) Beamed to Australia and New Zealand
(III) Beamed to Europe
(IV) Beamed to Asia
(a) Tuesday to Sunday
(b) Monday to Friday
(c) Wednesday and Saturday

INDIA

New Delhi (28°43′N, 77°12′E)

Ann: "This is All India Radio, General Overseas Service."
Int: Folk song played on a violin, cello, and tampura.

Stations:

3905 kHz, 100 kW	11770 kHz, 100 kW
6085 kHz, 100 kW	11775 kHz, 100 kW
7145 kHz, 100 kW	11810 kHz, 100 kW
7215 kHz, 100 kW	11825 kHz, 100 kW
7225 kHz, 100 kW	11850 kHz, 100 kW
9525 kHz, 100 kW	11880 kHz, 100 kW
9590 kHz, 100 kW	15080 kHz, 100 kW
9730 kHz, 100 kW	15165 kHz, 100 kW
9755 kHz, 100 kW	15190 kHz, 100 kW
9912 kHz, 100 kW	15205 kHz, 100 kW
11620 kHz, 100 kW	15335 kHz, 100 kW
11725 kHz, 100 kW	17387 kHz, 100 kW
11740 kHz, 100 kW	

Broadcasts:

Beamed to East and Southeast Asia

Time (UTC)	Freq (kHz)
1330–1500	11810, 15335
2245–0115	3905, 6085, 7215, 9590, 11770

Beamed to Northeast Asia

1000–1100	11725, 15190, 17387
2245–0115	9525, 11825, 11850

Beamed to Australia and New Zealand

1000–1100	11775, 15165, 15205
2045–2230	7145, 9912, 11740

Beamed to East Africa

1745–1945	9730, 15080

Beamed to West and North Africa

1945–2045	9755, 11880

Beamed to Western Europe and Great Britain

1745–2230	7225, 9525, 11620
1945–2230	9912

INDONESIA

Djakarta (6°12′S, 106°50′E)

Ann: "This is the Voice of Indonesia, broadcasting from Djakarta."
Int: Musical selection "Love Ambon" played on an electric organ.

Stations:

9710 kHz, 120 kW   11790 kHz, 120 kW

Broadcasts:

Time (UTC)	Freq (kHz)
0900–0930	9710, 11790
1100–1200	11790
2330–2400	9710, 11790

IRAN

Tehran (35°41′N, 51°27′E)

Ann: "This is Tehran, the Voice of Iran."
Int: A folk song played on a vibraphone.

Station:

9022 kHz, 250 kW

Broadcasts:

Time (UTC)	Freq (kHz)
2000–2030	9022

IRAQ

Baghdad (33°09′N, 44°35′E)

Ann: "This is Radio Baghdad."
Int: The sounds of a nightingale.

Station:

9758 kHz, 100 kW

Broadcasts:

Time (UTC)	Freq (kHz)
1930–2020	(I) 9758

Legend:

(I) Beamed to Europe

ISRAEL

Jerusalem (31°45′N, 35°15′E)

Ann: "This is Israel, broadcasting from Jerusalem."
Int: National anthem.

Stations:

5900 kHz, 100 kW	11655 kHz, 50 kW
7395 kHz, 50 kW	12045 kHz, 100 kW
7412 kHz, 50 kW	12055 kHz, 100 kW
9009 kHz, 50 kW	15100 kHz, 100 kW
9425 kHz, 50 kW	15465 kHz, 100 kW
9435 kHz, 50 kW	15485 kHz, 100 kW
9815 kHz, 50 kW	17815 kHz, 100 kW

Broadcasts:

Time (UTC)	Freq (kHz)
0500–0515	(I) 5900; (II) 7395, 9009; (III) 12045
1200–1230	(I) 11655, 12055, 15100, 15485; (III) 17815; (IV) 15465
2000–2030	(I) 5900, 7412; (V) 9009; (VI) 9425
2230–2300	(II) 5900, 7412, 9815; (VII) 9435

Legend:

  (I) Beamed to North America and Western Europe
 (II) Beamed to North America
(III) Beamed to Europe
(IV) Beamed to South and East Asia
 (V) Beamed to South Africa
(VI) Beamed to West Africa
(VII) Beamed to North America, Europe, France, North Africa and Central Africa

ITALY

Rome (41°48'N, 12°31'E)

Ann: "This is the Italian Radio and Television Service broadcasting from Rome."

Int: Birds chirping and chimes.

Stations:

5990 kHz, 100 kW	7290 kHz, 100 kW
6050 kHz, 100 kW	9575 kHz, 100 kW
7235 kHz, 100 kW	9710 kHz, 100 kW
7275 kHz, 100 kW	

Broadcasts:

Time (UTC)	Freq (kHz)
0100–0120	(I) 9575
0350–0410	(II) 6050, 7275, 9710
0415–0440	(III) 5990, 7275
2025–2045	(IV) 6050, 7235, 9575
2200–2225	(V) 5990, 9710

Legend:

(I) Beamed to North America
(II) Beamed to Europe
(III) Beamed to Mediterranean
(IV) Beamed to Near East
(V) Beamed to Australia and New Zealand

IVORY COAST

Abidjan (5°21'N, 3°75'W)

Ann: "This is the external service of Radio Abidjan."
Int: Clock chimes and the national anthem.

Station:

11920 kHz, 100 kW

Broadcasts:

Time (UTC)	Freq (kHz)
1830–2000	11920

JAPAN

Tokyo (36°10'N, 139°50'E)

Ann: "This is Radio Japan, the overseas broadcasting service of NHK."
Int: Musical notes played on a celesta, followed by "Kazoe Uta."

Stations:

9505 kHz, 100 kW	15195 kHz, 100 kW
9585 kHz, 100 kW	15235 kHz, 100 kW
9605 kHz, 100 kW	15270 kHz, 100 kW
9670 kHz, 200 kW	15300 kHz, 100 kW
9675 kHz, 100 kW	15310 kHz, 100 kW
9700 kHz, 200 kW	15325 kHz, 100 kW
11705 kHz, 100 kW	15420 kHz, 100 kW
11815 kHz, 100 kW	15430 kHz, 100 kW
11875 kHz, 100 kW	17725 kHz, 200 kW
11950 kHz, 100 kW	17825 kHz, 200 kW
15105 kHz, 100 kW	17880 kHz, 100 kW

Broadcasts:

General Services

Time (UTC)	Freq (kHz)
0000–0015	(I) 15105; (II) 9585; (III) 15195
0100–0115	(I) 15105; (II) 15310; (III) 17880
0200–0215	(I) 15105; (II) 15310; (III) 17880
0300–0315	(I) 15105; (II) 15310; (III) 17880
0400–0415	(I) 15105; (II) 15310; (III) 17880
0500–0515	(I) 15105; (II) 15310; (III) 17880
0600–0615	(I) 15105; (II) 15310; (III) 17880
0700–0715	(I) 9505; (II) 15310; (III) 17880
0800–0815	(I) 9505; (II) 15310; (III) 15195
0900–0915	(I) 9505; (II) 15310; (III) 15195
1000–1030	(I) 9505; (II) 9585; (III) 15195
1100–1115	(I) 9505; (II) 9585; (III) 15195
1200–1215	(I) 9505; (II) 9585; (III) 11815
1311–1315	(I) 9505; (II) 9585; (III) 11815
1400–1430	(I) 9505; (II) 9585; (III) 11815
1500–1515	(I) 9505; (II) 9585; (III) 11815
1611–1615	(I) 9505; (II) 9585; (III) 11815
1700–1715	(I) 9505; (II) 9585; (III) 11815
1800–1815	(I) 9505; (II) 9585; (III) 11815
1900–1915	(I) 15105; (II) 9585; (III) 11815
2000–2015	(I) 15105; (II) 9585; (III) 11815
2100–2115	(I) 15105; (II) 9585; (III) 11815
2200–2215	(I) 15105; (II) 9585; (III) 15195
2300–2330	(I) 15105; (II) 9585; (III) 15195

Regional Services

0130–0230	(IV) 15195, 15420, 17725, 17825
0800–0830	(II) 15325, 15430
0930–1030	(V) 11875, 15235
1115–1145	(VI) 9675, 11875
1630–1700	(VII) 9670, 11705
1830–1900	(II) 9605, 11950

| 2015–2045 | (VIII) 9700, 11875 |
| 2345–0045 | (I) 15270, 15300 |

Legend:

- (I) Beamed to North America
- (II) Beamed to Europe
- (III) Beamed to Asia
- (IV) Beamed to North America, Central America, South America, and Hawaii
- (V) Beamed to Australia and New Zealand
- (VI) Beamed to Southeast Asia
- (VII) Beamed to South Asia and Africa
- (VIII) Beamed to Middle East and North Africa

JORDAN

Amman (31°57′N, 35°56′E)

Ann: "This is Radio Jordan, broadcasting from Amman."
Int: Musical notes played on clarinet and piano.

Stations:

7155 kHz, 7.5 kW   9560 kHz, 100 kW

Broadcasts:

Time (UTC)	Freq (kHz)
1000–1300	7155
1500–1730	9560

KAMPUCHEA (Cambodia)

Phnom Penh (11°34′N, 104°51′E)

Ann: "Thini Phnom-Penh, Sathani Withayu Phsay Somleng Ronnacse Roup Roum Cheat Kampuchea."

Stations:

4907 kHz, 15 kW   6090 kHz, 50 kW

Broadcasts:

Time (UTC)	Freq (kHz)
0545–0600	4907
1245–1300	4907
2345–2400	4907, 6090

KENYA

Nairobi (1°30′S, 36°30′E)

Ann: "This is the voice of Kenya."

Stations:

4805 kHz, 5 kW   7120 kHz, 5 kW

Broadcasts:

Time (UTC)	Freq (kHz)
0255–0630	4805
0355–0630	(a) 4805
0645–1245	(c) 7120
0900–1100	7120
1300–2010	4805
1300–2110	(b) 4805

Legend:

(a) Sunday only
(b) Saturday only
(c) Saturday and Sunday only

KOREA (Democratic People's Republic; North)

Pyongyang (39°00′N, 126°00′E)

Ann: "This is Radio Pyongyang."

Stations:

3560 kHz, 50 kW	9420 kHz, 50 kW
3892 kHz, 50 kW	9510 kHz, 50 kW
6338 kHz, 50 kW	9768 kHz, 50 kW
6576 kHz, 50 kW	9820 kHz, 50 kW
7203 kHz, 50 kW	9977 kHz, 50 kW
7580 kHz, 50 kW	11535 kHz, 50 kW

Broadcasts:

Time (UTC)	Freq (kHz)
0500–0600	(I) (II) 3560, 9420, 9820
1200–1400	(III) 3560, 7580, 9510, 9768
2300–2400	(III) 3560, 9510, 11535
0600–0800	(III) 3560, 7580, 9820
1800–2000	(I) (II) (IV) 3560, 6338, 9977
1000–1100	(I) (V) 3560, 7203, 9768
2000–2130	(VI) 3892, 6576, 9420

Legend:

(I) Beamed to North America
(II) Beamed to Africa
(III) Beamed to Southeast Asia
(IV) Beamed to Central America
(V) Beamed to South America
(VI) Beamed to Europe

KOREA (Republic of; South)

Seoul (37°33′N, 126°55′E)

Ann: "This is Radio Korea, the overseas service of the Korean Broadcasting Corp., transmitting from Seoul."
Int: Korean folk song.

Stations:

7150 kHz, 50 kW	9640 kHz, 50 kW
7240 kHz, 50 kW	9665 kHz, 50 kW
9525 kHz, 50 kW	9675 kHz, 50 kW
9580 kHz, 50 kW	9720 kHz, 50 kW
9600 kHz, 50 kW	11860 kHz, 50 kW

Broadcasts:

Time (UTC)	Freq (kHz)
0300–0330	(I) 11860
0600–0630	(I) 9640, 9675
1000–1030	(I) (II) 9525, 9580
1130–1200	(II) 7150, 9665, 11860
1330–1400	(I) 9665, 11860
1600–1630	(I) 7240, 9640
1800–1830	(I) 9640, 9720
2000–2030	(III), (IV) 7150, 9600, 9720
2300–2330	(IV) 7150, 9600, 9640

Legend:

(I) Beamed to North America
(II) Beamed to South America
(III) Beamed to Europe
(IV) Beamed to Southeast Asia

KUWAIT

Kuwait (29°16′N, 47°53′E)

Ann: "This is Radio Kuwait."
Int: Kuwaiti folk song played on a clarinet.

Stations:

9555 kHz, 50 kW	11845 kHz, 250 kW
9580 kHz, 50 kW	15345 kHz, 250 kW

Broadcasts:

Time (UTC)	Freq (kHz)
0500–0800	(I) 15345
1700–2000	(I) 9555, 9580, 11845

Legend:

(I) Beamed to Europe and India

LESOTHO

Maseru (29°30'S, 27°30'E)

Ann: "This is Radio Lesotho."
Int: National Anthem.

Station:

4800 kHz, 10 kW

Broadcasts:

Time (UTC)	Freq (kHz)
0500–0530	4800
0730–0830	(a) 4800
1130–1300	4800
1500–1600	4800
1400–1630	(b) 4800

Legend:

(a) Weekdays only
(b) Sunday only

LIBERIA

Monrovia (6°14'N, 10°42'W)

Ann: "This is ELWA, broadcasting from Monrovia, Liberia."
Int: Tune played on Vibraharp.

Stations:

11945 kHz, 50 kW   11950 kHz, 50 kW

Broadcasts:

Time (UTC)	Freq (kHz)
0645–0730	(I) (a) 11950
0700–0730	(I) (b) 11950
1345–1615	(I) (b) 11945
1915–1945	(I) 11945
2015–2055	(I) 11945

Legend:

(I) Beamed to West and Central Africa
(a) Saturday only
(b) Sunday only

LUXEMBOURG

Luxembourg (49°45'N, 6°30'E)

Ann: "This is Radio Luxembourg."
Int: A popular Luxembourg song played on a piano.

Station:

6090 kHz, 500 kW

Broadcasts:

Time (UTC)	Freq (kHz)
0000–0246	6090

MALAGASY REPUBLIC

Tananarive (18°50′S, 47°35′E)

Ann: "This is Tananarive, the international service of Radio Malagasy."

Int: Folk song played on a valiha.

Station:

17730 kHz, 100 kW

Broadcasts:

Time (UTC)	Freq (kHz)
1500–1600	(I) 17730

Legend:

(I) Beamed to West and Central Africa

MALAYSIA

Kuala Lumpur (3°01′N, 101°46′E)

Ann: "This is the voice of Malaysia."

Int: National anthem.

Station:

15275 kHz, 100 kW

Broadcasts:

Time (UTC)	Freq (kHz)
0625–0855	15275

MALDIVES

Male (4°00′N, 74°00′E)

Ann: "This is the overseas service of Radio Maldives, broadcasting from Male."

Station:

4740 kHz, 30 kW

Broadcasts:

Time (UTC)	Freq (kHz)
1500–1730	4740

MARTINIQUE

Fort-de-France (14°30′N, 61°00′W)

Ann: "Ici Office de Radiodiffusion Television, Fort-de-France."
Int: A few bars of the selection "Adieu Foulards, Adieu Madras."

Station:

3315 kHz, 4 kW

Broadcasts:

Time (UTC)	Freq (kHz)
1000–0300	(a) 3315

Legend:

(a) French language only

MAURITANIA

Nouakchott (18°08′N, 16°00′W)

Ann: "Ici Nouakchott, Radiodiffusion Nationale de la Republique Islamique de Mauritanie."
Int: Musical notes played on a guitar.

Stations:

4850 kHz, 100 kW   9610 kHz, 4 kW
7245 kHz, 100 kW

Broadcasts:

Time (UTC)	Freq (kHz)
0700–0800	(a) 4850
1200–1430	(a) 7245, 9610
1800–2230	(a) 4850
0800–1400	(a) (b) 7245, 9610
1700–2300	(a) (c) 4850

Legend:

(a) French language
(b) Sundays only
(c) Saturdays and Sundays only

MAURITIUS

Port Louis (20°19′S, 57°31′E)

Ann: "This is the Mauritius Broadcasting Corporation."
Int: Musical selection "Extended Close."

Stations:

4850 kHz, 10 kW	9710 kHz, 10 kW

Broadcasts:

Time (UTC)	Freq (kHz)
0400–1300	9710
1300–1830	4850

MONACO

Monte Carlo (43°44′N, 7°26′E)

Ann: "This is Monte Carlo."
Int: Notes played on a music box.

Stations:

7105 kHz, 100 kW	9525 kHz, 100 kW
7245 kHz, 100 kW	9640 kHz, 100 kW

Broadcasts:

Time (UTC)	Freq (kHz)
0725–0900	(a) 7105
0900–1100	(b) 9525
0940–0955	(c) 9640
1030–1100	(d) 9525
1500–1515	(c) 7245

Legend:

(a) Daily
(b) Sunday only
(c) Special
(d) Monday through Thursday

MONGOLIAN PEOPLE'S REPUBLIC

Ulan Bator (48°00′N, 107°00′E)

Ann: "Ankarari Ulanbatras Jarsj Baina."

Stations:

5960 kHz, 50 kW	9574 kHz, 50 kW
6383 kHz, 50 kW	11856 kHz, 100 kW

Broadcasts:

Time (UTC)	Freq (kHz)		
1220–1250	(I)	(a)	5960, 6383
1715–1745	(I)	(a)	9574, 11856

Legend:

(I) Beamed to Southeast Asia
(a) Daily except Sunday

MOZAMBIQUE

Lourenco Marques (25°59′S, 32°22′E)

Ann: "This is Radio Mozambique."
Int: Drum beat.

Stations:

3210 kHz, 25 kW   6115 kHz, 25 kW
4865 kHz, 25 kW

Broadcasts:

Time (UTC)	Freq (kHz)
1800–1815	3210, 4865, 6115

NEPAL

Kathmandu (27°45′N, 85°20′E)

Ann: "This is Radio Nepal."
Int: Musical notes played on a conch shell, violin, piano, jal-tarang.

Stations:

3425 kHz, 100 kW   7105 kHz, 100 kW

Broadcasts:

Time (UTC)	Freq (kHz)
1435–1520	3425, 7105

NETHERLANDS

Hilversum (52°01′N, 5°02′E)

Ann: "This is Radio Nederland, the Dutch World Broadcasting System, in Hilversum."
Int: Dutch folk song played on a celesta.

Stations:

5955 kHz, 300 kW   6165 kHz, 300 kW
6020 kHz, 300 kW   7240 kHz, 100 kW
6045 kHz, 100 kW   9630 kHz, 100 kW

9660 kHz, 100 kW	11740 kHz, 300 kW
9715 kHz, 100 kW	15185 kHz, 100 kW
9895 kHz, 100 kW	17700 kHz, 100 kW
11730 kHz, 300 kW	17810 kHz, 100 kW

Broadcasts:

Time (UTC)	Freq (kHz)
0630–0750	(I) 9630
0800–0920	(II) 9715
0930–1050	(III) 5955, 6045, 7240, 9660
1400–1520	(IV) 9895, 11740, 15185, 17810
1400–1520	(V) 5955, 6045
1830–1950	(VI) 6020
1830–1950	(V) 6020
1830–1950	(VII) 11730, 17700
2000–2120	(V), (VII) 11730
2130–2250	(VIII) 9715, 11730
0200–0320	(VIII) 6165
0500–0620	(IX) 6165, 9715

Legend:

    (I) Beamed to New Zealand and Pacific Area
   (II) Beamed to Australia, New Zealand, and Pacific Area
  (III) Beamed to Europe
  (IV) Beamed to South and Southeast Asia
   (V) Beamed to Europe
  (VI) Beamed to Southern and Eastern Africa
 (VII) Beamed to West Africa
(VIII) Beamed to North America (East)
  (IX) Beamed to North America (West)

NEW HEBRIDES

Vila (17°44′S, 168°33′E)

Ann: "This is the New Hebrides."
Int: Native drums.

Stations:

3945 kHz, 2 kW	7260 kHz, 2 kW

Broadcasts:

Time (UTC)	Freq (kHz)
0030–0200	(a) 7260
0600–0900	(a) 3945

Legend:

(a) English and French languages

NEW ZEALAND

Wellington (41°05′S, 174°50′E)

Ann: "This is the external service of Radio New Zealand."
Int: The call of the New Zealand bellbird.

Stations:

6105 kHz, 7.5 kW	11780 kHz, 7.5 kW
9540 kHz, 7.5 kW	11960 kHz, 7.5 kW
9770 kHz, 7.5 kW	15130 kHz, 7.5 kW
11705 kHz, 7.5 kW	

Broadcasts:

Time (UTC)	Freq (kHz)
0100–0450	(I) 11705
0500–0700	(I) 9540
0500–1030	(I) 6105, 11780
0730–1030	(II) 6105
1800–2015	(I) 9770
1800–2130	(I) 11780
2030–0050	(I) 11960
2030–0450	(I) 15130

Legend:

(I) Beamed to Pacific Area
(II) Beamed to Australia

NIGERIA

Ibadan (7°23′N, 3°55′E)

Ann: "This is the Western Nigeria Broadcasting Service."
Int: Two chimes.

Stations:

7275 kHz, 100 kW	15185 kHz, 100 kW
15120 kHz, 100 kW	

Broadcasts:

Time (UTC)	Freq (kHz)
0645–0835	(I) (II) 7275; (III) 15120
1530–1700	(I) 7275; (II) 15120; (IV) 15185
1800–1930	(I) 7275; (IV) 15185

Legend:

(I) Beamed to West Africa
(II) Beamed to East Africa and Middle East
(III) Beamed to North Africa, Mediterranean, and Europe
(IV) Beamed to Central and South Africa

NORWAY

Oslo (59°11′N, 10°58′E)

Ann: "This is the overseas service of Radio Norway."
Int: Ancient folk song.

Stations:

6015 kHz, 100 kW	11860 kHz, 250 kW
6180 kHz, 250 kW	11895 kHz, 250 kW
9550 kHz, 100 kW	15135 kHz, 100 kW
9605 kHz, 250 kW	15175 kHz, 100 kW
9610 kHz, 250 kW	15345 kHz, 250 kW
9645 kHz, 250 kW	17840 kHz, 250 kW
11850 kHz, 100 kW	21730 kHz, 250 kW

Broadcasts:

The following broadcasts are made on Sunday only:

Time (UTC)	Freq (kHz)
0800–0830	(I) 11850; (II) 11895, 15135
1200–1230	(III) 6015; (IV) 15345; (V) 21730
1400–1430	(III) 9605; (VI) 17840; (VII) 21730
1600–1630	(VIII) 15175, 15345; (IX) 21730
1800–1830	(VII) 11860, 15345; (VIII) 15175
2000–2030	(VII) 15175; (X) 9610, 11860
2200–2230	(VI) 9645, 15175; (XI) 11860

The following broadcasts are made on Monday only:

0000–0030	(VI) 6180, 9550; (XII) 9605
0200–0230	(VI) 6180, 9550, 9645
0400–0430	(V) 9645; (VI) 9550; (X) 6180
0600–0630	(V) 11850, 15175; (X) 9645

Legend:

- (I) Beamed to Australia and New Zealand
- (II) Beamed to Far East
- (III) Beamed to Europe
- (IV) Beamed to Indonesia
- (V) Beamed to Middle East
- (VI) Beamed to North America
- (VII) Beamed to Africa
- (VIII) Beamed to North America and Central America
- (IX) Beamed to South America
- (X) Beamed to Pacific Ocean Area
- (XI) Beamed to West Indies and South America
- (XII) Beamed to Far East

PAKISTAN

Karachi (24°55′N, 67°00′E)

Ann: "This is Radio Pakistan."
Int: Folk song.

Stations:

6115 kHz, 50 kW	15110 kHz, 50 kW
7085 kHz, 50 kW	15325 kHz, 50 kW
9460 kHz, 50 kW	17665 kHz, 50 kW
11675 kHz, 50 kW	17830 kHz, 50 kW
11885 kHz, 50 kW	21590 kHz, 50 kW

Broadcasts:

Time (UTC)	Freq (kHz)
0230–0245	(I) 17830, 21590
0430–0445	(II) 11885, 15325, 17830
1100–1115	(III) 15110, 17665
1630–1645	(IV) 9460, 11675
2100–2145	(III) 6115, 7085

Legend:

(I) Beamed to Far East
(II) Beamed to East Africa
(III) Beamed to Europe
(IV) Beamed to Near and Middle East

PANAMA

David (8°30′N, 82°30′W)

Station:

6045 kHz, 1 kW

Broadcasts:

Time (UTC)	Freq (kHz)
0000–0500	(a) 6045
1100–2400	(a) 6045

Legend:

(a) Spanish language only

PAPUA NEW GUINEA

Port Moresby (9°30′S, 147°30′E)

Ann: "This is Radio Port Moresby."

Stations:

4890 kHz, 10 kW	9520 kHz, 10 kW

Broadcasts:

Time (UTC)	Freq (kHz)
0530–1400	4890
2200–0800	9520

PHILIPPINES

Manila (14°48′N, 120°55′E)

Ann: "This is the English service of the Far East Broadcasting Company, transmitting from the Philippines."
Int: Musical selection "We have heard the Joyful Sound."

Stations:

7225 kHz, 50 kW	15390 kHz, 50 kW
9505 kHz, 50 kW	15440 kHz, 50 kW
11890 kHz, 50 kW	17810 kHz, 50 kW
11920 kHz, 50 kW	21515 kHz, 50 kW

Broadcasts:

Time (UTC)	Freq (kHz)
0100–0600	(I) (II) 15390, 17810
0600–0800	(I) (II) 15390
0800–0900	(I) 9505
0800–1000	(III) 11890, 11920
1100–1130	(IV) 7225, 11890
1215–1630	(I) (II) 15440
1400–1630	(I) (II) 11920
2300–0100	(I) (II) 11890
2300–0230	(III) 21515
2330–2400	(IV) 7225, 11890
2345–0100	(IV) 17810

Legend:

(I) Beamed to Cambodia, Thailand, Vietnam, and Burma
(II) Beamed to USSR
(III) Beamed to Australia, New Zealand, and New Guinea
(IV) Beamed to Japan

Manila

Ann: "Radio Philippines, Voice of the Philippines, National Media Production Center."

Stations:

11725 kHz, 50 kW	15280 kHz, 50 kW
11875 kHz, 50 kW	

Broadcasts:

Time (UTC)	Freq (kHz)
0100–0200	11725, 15280
1400–1500	11725, 11875

POLAND

Warsaw (52°04′N, 20°52′E)

Ann: "This is Polskie Radio Warsaw."
Int: National anthem.

Stations:

6135 kHz, 100 kW	9675 kHz, 100 kW
6155 kHz, 100 kW	11800 kHz, 100 kW
7270 kHz, 40 kW	11815 kHz, 100 kW
7285 kHz, 40 kW	11840 kHz, 100 kW
9525 kHz, 100 kW	15120 kHz, 100 kW
9540 kHz, 40 kW	15275 kHz, 100 kW

Broadcasts:

Time (UTC)	Freq (kHz)
0200–0400	(I) 6135, 7270, 9675, 11800, 15275
0630–0700	(II) 6155, 7285, 9540
1200–1230	(II) 6155, 7270
1600–1630	(II) 6135, 9525
1830–1900	(II) 6135, 9525
2030–2100	(II) 6135, 9525
2230–2300	(II) 6135, 9540
1230–1300	(III) 9675, 11815, 15120
1630–1700	(III) 7270, 9525, 11840
2000–2030	(III) 7270, 9525, 11800

Legend:

(I) Beamed to North America
(II) Beamed to Europe
(III) Beamed to Africa

PORTUGAL

Lisbon (38°45′N, 8°40′W)

Ann: "This is Radio Portugal, speaking to you from Lisbon."
Int: A few bars from the selection "Fadango."

Stations:

6025 kHz, 100 kW	17880 kHz, 100 kW
9740 kHz, 100 kW	17895 kHz, 100 kW
15340 kHz, 100 kW	17935 kHz, 100 kW

Broadcasts:

Time (UTC)	Freq (kHz)
0230–0300	(I) 6025, 17935
0500–0530	(I) 6025, 17935
1400–1430	(II) (a) 17895
1600–1630	(II) (a) 17895
1700–1730	(III) (b) 15340
1800–1830	(III) (a) 17880
1830–1900	(IV) 6025, 9740

Legend:

- (I) Beamed to North America
- (II) Beamed to India and Middle East
- (III) Beamed to Africa
- (IV) Beamed to Europe
- (a) Daily except Sunday
- (b) Sunday only

RHODESIA

Salisbury (18°00′S, 31°30′E)
Ann: "This is the Rhodesia Broadcasting Corp."
Int: Church bells.

Stations:

2425 kHz, 20 kW   5012 kHz, 100 kW
3396 kHz, 100 kW

Broadcasts:

Time (UTC)	Freq (kHz)
0355–0430	(a) 2425
1700–2200	2425
1700–2100	(b) 2425
0355–0545	(a) 3396
1700–2200	3396
1700–2100	(b) 3396
0545–1630	5012
0500–1545	(b) 5012

Legend:

- (a) Monday through Saturday
- (b) Sunday only

ROMANIA

Bucharest (44°25′N, 26°06′E)

Ann: "You are tuned to Radio Bucharest."
Int: Folk song.

Stations:

5990 kHz, 120 kW	9690 kHz, 120 kW
6150 kHz, 120 kW	11790 kHz, 240 kW
6190 kHz, 120 kW	11830 kHz, 240 kW
7195 kHz, 120 kW	11840 kHz, 240 kW
7225 kHz, 120 kW	11940 kHz, 240 kW
9510 kHz, 120 kW	15250 kHz, 240 kW
9530 kHz, 120 kW	15340 kHz, 120 kW
9540 kHz, 120 kW	15345 kHz, 120 kW
9570 kHz, 120 kW	15365 kHz, 120 kW
9685 kHz, 120 kW	15380 kHz, 120 kW

Broadcasts:

Time (UTC)	Freq (kHz)
0130–0230	(I) 5990, 6190, 9510
0400–0430	(I) 5990, 9570, 11940
0530–0600	(II) 9685, 11840, 15340
0645–0715	(III) 11830, 11940, 15250, 15380
1200–1230	(IV) 11830, 15345
1300–1330	(V) 9690, 11940, 15250
1500–1530	(IV) 7225, 9530, 11940
1730–1800	(II) 9540, 11790, 15365
1930–2030	(V) 6150, 7195
2100–2130	(V) 5990, 7225

Legend:

  (I) Beamed to North America
  (II) Beamed to Africa
  (III) Beamed to Pacific Area
  (IV) Beamed to Asia
  (V) Beamed to Europe

SABAH

Kota Kinabalu (6°12′N, 116°14′E)

Ann: "Inilah Radio Malaysia, Kota Kinabalu."
Int: National anthem.

Station:

  4970 kHz, 10 kW

Broadcasts:

Time (UTC)	Freq (kHz)
0500–0700	4970
1200–0200	4970

SAO TOME AND PRINCIPE

Sao Tome (0°00′, 6°30′E)

Ann: "Aqui Republica Democratica de Sao Tome e Principe, falavos Radio Nacional."

Station:

4807 kHz, 1 kW

Broadcasts:

Time (UTC)	Freq (kHz)
0530–2300	(a) 4807

Legend:

(a) Portuguese language only

SARAWAK

Kuching (1°33′N, 110°20′E)

Ann: "Inilah Radio Malaysia, Sarawak."
Int: Folk song.

Stations:

4950 kHz, 10 kW	9605 kHz, 10 kW
7160 kHz, 10 kW	

Broadcasts:

Time (UTC)	Freq (kHz)
0800–1600	4950
0900–1600	9605
2200–0100	4950
2200–0730	9605
2300–1445	7160

SAUDI ARABIA

Riyadh (24°30′N, 46°23′E)

Ann: "This is the broadcasting service of the Kingdom of Saudi Arabia."
Int: Call to prayer played on a flute.

Station:

11855 kHz, 50 kW

Broadcasts:

Time (UTC)	Freq (kHz)
1000–1300	(I) 11855
1900–2200	(I) 11855

SENEGAL

Dakar (14°39′N, 17°26′W)

Ann: "This is the English program of the international service of the Senegalese Broadcasting System."
Int: Music played on a harp.

Stations:

4890 kHz, 25 kW   11895 kHz, 100 kW

Broadcasts:

Time (UTC)	Freq (kHz)
1840–1900	4890, 11895

SEYCHELLES

Victoria (4°36′S, 55°28′E)

Ann: "This is the FEBA, Seychelles."

Stations:

11870 kHz, 100 kW   15160 kHz, 100 kW

Broadcasts:

Time (UTC)	Freq (kHz)
0600–0830	15160
1530–1600	11870

SIERRA LEONE

Freetown (8°30′N, 13°14′W)

Ann: "This is the Sierra Leone Broadcasting Service."
Int: Military tune.

Station:

3316 kHz, 10 kW

Broadcasts:

Time (UTC)	Freq (kHz)
0600–1015	3316

SINGAPORE

Singapore (1°20′N, 103°42′E)

Ann: "This is Radio Singapore."

5010 kHz, 20 kW   11940 kHz, 50 kW

Broadcasts:

Time (UTC)	Freq (kHz)
0000–1630	5010, 11940
2230–1630	5010, 11940

SOLOMON ISLANDS

Honiara (9°35′S, 160°03′E)

Ann: "This is the Solomons Radio."
Int: Drums and bamboo pipes.

Stations:

5015 kHz, 5 kW   9545 kHz, 5 kW

Broadcasts:

Time (UTC)	Freq (kHz)
0600–1130	5015
1900–2200	5015
2130–0300	9545

SOMALIA

Mogadiscio (2°06′N, 45°06′E)

Ann: "This is the voice of the Somali Democratic Republic."
Int: Musical notes played on a guitar.

Station:

9585 kHz, 25 kW

Broadcasts:

Time (UTC)	Freq (kHz)
1100–1130	9585

SOUTH AFRICA

Johannesburg (26°35′S, 28°08′E)

Ann: "This is Radio RSA, the Voice of South Africa, calling from
      Johannesburg."
Int: The call of the bokmakierie bird.

Stations:

4875 kHz, 100 kW	11900 kHz, 100 kW
5980 kHz, 100 kW	15220 kHz, 100 kW
7270 kHz, 100 kW	17780 kHz, 100 kW
9585 kHz, 100 kW	21535 kHz, 100 kW

Broadcasts:

Time (UTC)	Freq (kHz)	
0300–0430	(I)	4875, 5980, 7270
0600–0700	(II)	11900, 15220, 17780
1100–1200	(III)	11900, 15220, 21535
1300–1600	(III)	11900, 15220, 21535
1600–1700	(IV)	4875, 11900, 21535
2100–2200	(V)	4875, 5980, 9585
2230–2330	(VI)	5980, 9585, 11900

Legend:

(I) Beamed to East Africa
(II) Beamed to West Africa
(III) Beamed to Central and East Africa
(IV) Beamed to East Africa and Middle East
(V) Beamed to West Africa and Europe
(VI) Beamed to North America

SPAIN

Madrid (40°18′N, 3°31′W)

Ann: "This is the voice of Spain, broadcasting from the Spanish National Radio."
Int: Ringing of a gong.

Stations:

6065 kHz, 350 kW		9505 kHz, 100 kW	
6100 kHz, 350 kW		11880 kHz, 100 kW	

Broadcasts:

Time (UTC)	Freq (kHz)	
0100–0400	(I)	6065, 11880
2030–2230	(II)	6100,9505

Legend:

(I) Beamed to North and South America
(II) Beamed to Europe

SRI LANKA

Colombo (7°06′N, 79°54′E)

Ann: "This is the Sri Lanka Broadcasting Corp."
Int: Oriental folk song.

Stations:

4940 kHz, 10 kW		6005 kHz, 10 kW
4970 kHz, 10 kW		6130 kHz, 10 kW

Broadcasts:

Time (UTC)	Freq (kHz)
0030–0300	(a) 4940
0300–0430	6130
0430–0730	(a) 6005
0730–1030	6130
1030–1730	(a) 4940
1330–1730	(b) 4970

Legend:

(a) Saturday and Sunday only
(b) Monday through Friday

SWEDEN

Stockholm (59°30′N, 18°00′E)

Ann: "This is Radio Sweden."
Int: The first few bars of "Storm och boljor tystna re'n."

Stations:

6065 kHz, 100 kW	11905 kHz, 100 kW
6120 kHz, 100 kW	15120 kHz, 100 kW
9605 kHz, 100 kW	15240 kHz, 100 kW
9630 kHz, 100 kW	15305 kHz, 100 kW
9695 kHz, 100 kW	17730 kHz, 100 kW
11705 kHz, 100 kW	17795 kHz, 100 kW
11845 kHz, 100 kW	21690 kHz, 100 kW

Broadcasts:

Time (UTC)	Freq (kHz)
0030–0100	(I) 11905
0230–0300	(I) 9695, 11705
1100–1130	(II) 9630; (III) 15305; (IV) 21690
1230–1300	(V) 15120; (I) 15305; (IV) 21690
1400–1430	(VI) 15240, 17795; (I) 15305
1600–1630	(II) 6065; (VII) 15240
1830–1900	(II) 6065; (IV) 15240, 17730
2100–2130	(II) 6065; (IV) 11845; (VII) 9605
2300–2330	(I) 6120, 9695, 11705

Legend:

  (I) Beamed to North America
 (II) Beamed to Europe
(III) Beamed to Australia and New Zealand
(IV) Beamed to Africa
 (V) Beamed to East Asia
(VI) Beamed to South Asia
(VII) Beamed to Middle East

SWITZERLAND

Bern (46°49'N, 7°24'E)

Ann: "This is Radio Switzerland."
Int: A Swiss folk song played on a music box.

Stations:

5965 kHz, 100 kW	11905 kHz, 100 kW
6135 kHz, 100 kW	15140 kHz, 100 kW
9590 kHz, 100 kW	15305 kHz, 100 kW
9725 kHz, 100 kW	15430 kHz, 100 kW
11715 kHz, 100 kW	17830 kHz, 100 kW
11720 kHz, 100 kW	17840 kHz, 100 kW
11775 kHz, 100 kW	21520 kHz, 100 kW
11870 kHz, 100 kW	

Broadcasts:

Beamed to North America (West) and Central America

Time (UTC)	Freq (kHz)
0145–0215	5965, 6135, 9725, 11715
0430–0500	9725, 11715
1315–1345	15140

Beamed to East Asia, Far East, India, Southeast Asia, Australia, New Zealand

0700–0730	9590, 11775, 15305, 17840
0900–0930	9590, 11775, 15305, 17840
1315–1345	11775, 11905, 15430, 17830

Beamed to Middle East

1530–1600	11870, 15430, 17830

Beamed to Africa

1100–1130	15140, 15430, 17830, 21520
1530–1600	11870, 15430, 17830
2100–2130	9590, 11720, 11870, 15305

SYRIAN ARAB REPUBLIC

Damascus (33°30'N, 36°07'E)

Ann: "This is the Syrian Arab Republic Broadcasting Service from Damascus."
Int: Folk song played on a guitar.

Stations:

7105 kHz, 50 kW	9545 kHz, 50 kW

Broadcasts:

Time (UTC)	Freq (kHz)
0615–0625	(I) 7105
0700–0730	(II) 7105
2030–2200	(II) 9545

Legend:

(I) Beamed to Israel
(II) Beamed to West Turkey

TAHITI

Papeete (17°00′S, 149°00′W)

Ann: "Ici Tahiti Office de Radiodiffusion Television Francaise."
Int: Flute and drums.

Stations:

6135 kHz, 4 kW	11825 kHz, 20 kW
9750 kHz, 4 kW	15170 kHz, 20 kW

Broadcasts:

Time (UTC)	Freq (kHz)
1900–1915	6135, 9750, 11825, 15170

TAIWAN (Formosa; Nationalist China)

Taipei (25°09′N, 121°24′E)

Ann: "This is the voice of Free China."

Stations:

9510 kHz, 100 kW	15225 kHz, 100 kW
9600 kHz, 100 kW	15345 kHz, 100 kW
9685 kHz, 100 kW	15425 kHz, 100 kW
9765 kHz, 100 kW	17720 kHz, 100 kW
11825 kHz, 100 kW	17890 kHz, 100 kW
11860 kHz, 100 kW	

Broadcasts:

Time (UTC)	Freq (kHz)
0100–0200	(I) 9765, 15345, 15425, 17890
0100–0200	(II) 11825
0300–0350	(I) 15345, 17890
0300–0350	(II) 11825
1700–1800	(I) 9685, 17890
1700–1800	(II) 11825
2130–2230	(III) 9510, 9600, 11860, 15225, 17720

Legend:

(I) Beamed to North America
(II) Beamed to Australia, New Zealand
(III) Beamed to Africa, Europe

TANZANIA

Dar es Salaam (6°50′S, 39°14′E)

Ann: "This is the external service of Radio Tanzania, Dar es Salaam."
Int: Song played on a celesta.

Stations:

4785 kHz, 10 kW	9750 kHz, 100 kW
6105 kHz, 100 kW	15435 kHz, 50 kW

Broadcasts:

Time (UTC)	Freq (kHz)
0330–0430	4785
0430–0530	6105
0900–1030	9750
1600–1700	4785
1700–1830	4785, 15435
1830–2000	15435

THAILAND

Bangkok (14°00′N, 100°30′E)

Ann: "This is Radio Thailand, Bangkok."

Stations:

9655 kHz, 2.5 kW	11905 kHz, 100 kW

Broadcasts:

Time (UTC)	Freq (kHz)
2330–0155	(I) 9655
0415–0515	(I) 9655, 11905
1055–1155	(II) 9655, 11905

Legend:

(I) Beamed to North America
(II) Beamed to Southeast Asia

TOGO

Lome (6°16′N, 1°12′E)

Ann: "Ici Lome, Radiodiffusion du Togo."
Int: The hymn "Togolais."

Stations:

5047 kHz, 100 kW   6155 kHz, 100 kW

Broadcasts:

Time (UTC)	Freq (kHz)
1245–1300	5047, 6155
1950–2000	5047, 6155

TURKEY

Ankara (39°54′N, 30°42′E)

Ann: "This is the Voice of Turkey."
Int: Folk song.

Stations:

9515 kHz, 100 kW   11880 kHz, 250 kW

Broadcasts:

Time (UTC)	Freq (kHz)
2200–0030	9515, 11880

UGANDA

Kampala (0°20′N, 32°36′E)

Ann: "This is Uganda Broadcasting Corp., Kampala."
Int: African drums.

Stations:

6030 kHz, 250 kW   9730 kHz, 250 kW
9515 kHz, 250 kW   15325 kHz, 250 kW

Broadcasts:

Time (UTC)	Freq (kHz)
1430–1530	(I) 6030
1615–1730	(II) 9515
1800–1830	(III) 15325
2030–2100	(IV) 9730

Legend:

(I) Beamed to East and Central Africa
(II) Beamed to South Africa
(III) Beamed to West Africa
(IV) Beamed to North Africa, Middle East, and Europe

UNITED NATIONS

New York, N.Y., USA (40°45′N, 74°00′W)

Ann: "This is the United Nations."
Int: Musical chimes.

Stations:

5955 kHz, 250 kW	15155 kHz, 250 kW
6055 kHz, 250 kW	15285 kHz, 250 kW
6090 kHz, 250 kW	15305 kHz, 250 kW
9630 kHz, 250 kW	15410 kHz, 250 kW
9650 kHz, 250 kW	15415 kHz, 250 kW
9710 kHz, 250 kW	17750 kHz, 250 kW
11825 kHz, 250 kW	17865 kHz, 250 kW
11930 kHz, 250 kW	21670 kHz, 250 kW

Broadcasts:

Time (UTC)	Freq (kHz)
1800–1805	(I) (a) 15305, 15410, 21670
1830–1835	(II) (a) 15305, 15410, 21670
0545–0600	(III) (a) 6055
0630–0635	(IV) (a) 9630, 11825, 15285
0700–0707	(V) (a) 6055, 9710
0230–0245	(VI) (a) 15155, 17750
0845–0900	(VII) (a) 5955, 9650
1000–1005	(VIII) (a) 11930

Broadcasts of Security Council Meetings When in Session

1430–1800	(IX) 15410
1430–1800	(X) (b) 15410, 21670
1900–2200	(IX) 15410
1900–2300	(IX) 17865
2200–2400	(IX) 15415
2300–0200	(IX) 6090

Legend:

- (I) Beamed to Central Europe
- (II) Beamed to West and Central Africa
- (III) Beamed to North, West, and Central Africa
- (IV) Beamed to Europe, Middle East, and Africa
- (V) Beamed to West and Central Africa
- (VI) Beamed to East Asia, Southeast Asia, and Malaysia
- (VII) Beamed to Australia and New Zealand
- (VIII) Beamed to East Asia, South Asia, and Northeast Asia
- (IX) Beamed to Africa
- (X) Beamed to Middle East
- (a) Saturday only
- (b) English and French language

USA

Scituate, Mass. (42°13'N, 70°44'W) (Studios in Oakland, Calif.)
(Formerly Radio New York Worldwide)

Ann: "This is WYFR, Oakland, Calif. Family Radio Network."

Station:

6155 kHz, 100 kW

Broadcasts:

Beamed to Mexico

Time (UTC)	Freq (kHz)
0100–0135	6155
0200–0230	6155
0300–0430	6155

Red Lion, Pa. (39°50'N, 76°34'W)

Ann: "You are listening to World International Broadcasters, WINB, Red Lion."

Stations:

15185 kHz, 50 kW
15305 kHz, 50 kW
17875 kHz, 50 kW

Broadcasts:

Beamed to Western Europe, Mediterranean, North Africa

Time (UTC)	Freq (kHz)
1700–1930	17875
1930–2100	15305
2100–2245	15185

Washington, DC (38°50'N, 77°00'W)
(AFRTS)

Ann: "This is the American Forces Radio and Television Service."

Stations:

Some of the stations listed below are located in the USA; others are located in various countries throughout the world, relaying broadcasts from Washington.

5995 kHz, 100 kW	11805 kHz, 100 kW
6030 kHz, 100 kW	11840 kHz, 100 kW
6095 kHz, 100 kW	15330 kHz, 100 kW
9700 kHz, 100 kW	15430 kHz, 100 kW
9755 kHz, 100 kW	17765 kHz, 100 kW
11790 kHz, 100 kW	

Broadcasts:

Beamed to Far East and Pacific Area

Time (UTC)	Freq (kHz)
0300–1200	11840
0400–0800	11805
0500–1630	9700
0800–1500	6095
1200–0030	5995
1500–1630	11805
2130–0400	17765
2130–0500	15330

Beamed to Europe

1330–2300	15430
1730–2300	15330
1800–2300	11790

Beamed to Caribbean and Antarctica

0030–1130	6030
1600–1300	9755
1230–2330	15330
2300–0130	9755

Washington, DC
(VOA)

Ann: "This is the Voice of America."
Int: Musical selection "Yankee Doodle."

Stations:

Transmitting and relay stations are located in Greenville, North Carolina; Bethany, Ohio; Delano, California; Dixon, California; Marathon, Florida; and various countries throughout the world.

3980 kHz, 25 kW		6185 kHz, 50 kW
3990 kHz, 25 kW		7110 kHz, 50 kW
5955 kHz, 25 kW		7165 kHz, 50 kW
5995 kHz, 25 kW		7170 kHz, 50 kW
6010 kHz, 250 kW		7195 kHz, 50 kW
6015 kHz, 250 kW		7200 kHz, 50 kW
6035 kHz, 50 kW		7230 kHz, 50 kW
6040 kHz, 50 kW		7270 kHz, 50 kW
6045 kHz, 50 kW		7280 kHz, 50 kW
6060 kHz, 50 kW		7295 kHz, 50 kW
6080 kHz, 50 kW		7325 kHz, 50 kW
6095 kHz, 50 kW		9515 kHz, 250 kW
6110 kHz, 50 kW		9530 kHz, 250 kW
6130 kHz, 50 kW		9545 kHz, 250 kW
6140 kHz, 50 kW		9565 kHz, 100 kW

9640 kHz, 100 kW	15185 kHz, 250 kW
9670 kHz, 100 kW	15205 kHz, 250 kW
9680 kHz, 100 kW	15220 kHz, 500 kW
9700 kHz, 100 kW	15250 kHz, 500 kW
9705 kHz, 100 kW	15290 kHz, 250 kW
9730 kHz, 250 kW	15430 kHz, 250 kW
9740 kHz, 250 kW	17710 kHz, 250 kW
9760 kHz, 50 kW	17785 kHz, 250 kW
9770 kHz, 50 kW	17790 kHz, 250 kW
11715 kHz, 50 kW	17820 kHz, 100 kW
11740 kHz, 250 kW	17870 kHz, 100 kW
11760 kHz, 250 kW	17895 kHz, 250 kW
11820 kHz, 250 kW	21485 kHz, 100 kW
11925 kHz, 250 kW	21610 kHz, 100 kW
11935 kHz, 100 kW	21670 kHz, 100 kW
15150 kHz, 250 kW	

Broadcasts:

Beamed to Europe

Time (UTC)	Freq (kHz)
0300–0400	7200, 7325
0300–0500	6040
0300–0600	6060
0400–0600	5995
0400–0700	9705
0500–0700	9670
0600–0700	7230, 7295
1600–1900	3980
1700–1715	6040
1700–1730	17785
1800–2200	15205
1800–1930	9760
1830–2200	7170
1900–2200	15250
2100–2200	11760

Beamed to North Africa

0300–0700	5955
0500–0700	7200
0500–0730	5995
0600–0700	9670
1600–1830	21670
1800–2100	17785
1900–2100	6040
2000–2200	9760

Beamed to East, South, and West Africa

0300–0400	9680
0300–0530	11925

```
0300-0600    9530
0300-0730    3990, 6035, 6080, 7195, 7280
0400-0730    9740
0600-0730    15430
1600-2200    3990, 6045, 7195, 9515, 21485
1830-2200    17710
1900-2200    17870
```

Beamed to Middle East

```
0300-0400    9770
0300-0430    7200
0600-0630    7270
1700-1800    7170
1700-2200    6015, 6140
1900-2200    9700
2100-2200    9760
```

Beamed to South Asia

```
0100-0300    7110, 11740, 11760, 15185, 15220, 17790
1300-1500    11935
1300-1600    6110
1430-1600    15150
1500-1600    11715
1600-1800    6140, 7110, 11820, 15150
```

Beamed to Northeast Asia

```
0100-0200    17895
1100-1500    6010, 6110, 6185, 7165, 9565
2200-0100    6185, 15250, 15290
```

Beamed to Southeast Asia

```
1100-1500    6110
1130-1200    11715
1430-1600    9760
2200-0100    6185, 9545
```

Beamed to Indonesia

```
1100-1500    7165
2200-2300    11760
2300-2400    6095
```

Beamed to Pacific Area

```
1100-2400    5955, 9730
2200-2400    15290, 17820, 21610
```

Beamed to Philippines and Malaysia

```
1100-1500    7165
```

Beamed to South America

```
0000-0200    6130, 9640, 11740, 15205
```

Kiev (50°20′N, 30°30′E)

Ann: "This is Radio Kiev."
Int: Musical tones played on a celesta.

Stations:

5900 kHz, 120 kW		7215 kHz, 120 kW	
5920 kHz, 120 kW		7245 kHz, 120 kW	
6020 kHz, 120 kW		9610 kHz, 120 kW	
6170 kHz, 120 kW		11690 kHz, 120 kW	
7150 kHz, 120 kW		15100 kHz, 120 kW	

Broadcasts:

Time (UTC)	Freq (kHz)
0030–0100	(I) 5900, 7150, 7215, 9610, 11690, 15100
0300–0330	(I) 7215, 7245
1930–2000	(II) 5920, 6020, 6170

Legend:

(I) Beamed to North America
(II) Beamed to Europe

Moscow (55°45′N, 37°30′E)

Ann: "This is Radio Moscow."
Int: Musical notes played on a celesta.

Stations:

6070 kHz, 120 kW	9710 kHz, 120 kW
6150 kHz, 120 kW	9720 kHz, 120 kW
7100 kHz, 120 kW	9760 kHz, 120 kW
7130 kHz, 120 kW	9780 kHz, 120 kW
7160 kHz, 120 kW	11690 kHz, 120 kW
7170 kHz, 120 kW	11700 kHz, 120 kW
7200 kHz, 120 kW	11720 kHz, 120 kW
7250 kHz, 120 kW	11740 kHz, 120 kW
7290 kHz, 120 kW	11750 kHz, 120 kW
7310 kHz, 120 kW	11800 kHz, 120 kW
7350 kHz, 120 kW	11830 kHz, 120 kW
7390 kHz, 120 kW	11850 kHz, 120 kW
9450 kHz, 120 kW	11870 kHz, 120 kW
9530 kHz, 120 kW	11960 kHz, 120 kW
9550 kHz, 120 kW	12000 kHz, 120 kW
9555 kHz, 120 kW	12020 kHz, 120 kW
9610 kHz, 120 kW	12050 kHz, 120 kW
9650 kHz, 120 kW	15100 kHz, 120 kW
9660 kHz, 120 kW	15130 kHz, 120 kW
9680 kHz, 120 kW	15180 kHz, 120 kW
9700 kHz, 120 kW	15210 kHz, 120 kW

```
15250 kHz,  120 kW          15420 kHz,  120 kW
15300 kHz,  120 kW          17700 kHz,  120 kW
15380 kHz,  120 kW          21510 kHz,  120 kW
```

Broadcasts:

Beamed to South Asia

Time (UTC)	Freq (kHz)
1000–1130	9650, 9680, 11850, 11960, 15050, 15210
1500–1530	9650, 9700, 12020, 11850
1600–1630	9660, 11850, 15100, 15210

Beamed to Africa

0345–0400	7100, 7160, 9660, 9700, 11870, 11960, 15210, 15420
1330–1400	15100, 15420, 17700, 21510
2000–2300	6070, 6150, 9780, 11870

Beamed to Europe

1230–1330	9450, 9720, 11700, 11740, 11830, 15300
2000–2030	7200, 7250, 7310, 7390, 9550, 9720, 11830
2100–2130	7200, 7250, 7310, 7390, 9550, 9720
2200–2300	7250, 7390, 9555, 9610, 9720, 9760, 11800
2300–2330	7390, 9610, 9760, 11800

Beamed to North America (Pacific Coast)

0330–0730	9610, 9710, 11690, 11720, 11750, 12000, 12020, 12050, 15100, 15180

Beamed to North America (East Coast)

0000–0200	7100, 7130, 7160, 7170, 7250, 7350, 9530, 9660, 9680, 9700, 11850, 11870, 11960, 12020, 12050, 15100, 15210, 15420
0200–0400	6070, 7100, 7130, 7160, 7170, 7250, 7290, 9530, 9650, 9680, 9700, 11850, 11870, 11960, 12020, 12050, 15100, 15210, 15420
2230–2400	7130, 7160, 7170, 7250, 7350, 9530, 9650, 9660, 9680, 9720, 11750, 11850, 11870, 11960, 12020, 12050, 15100, 15210, 15420

Beamed to Australia and New Zealand

1800–1830	9780, 11870, 15130, 15250, 15380
1830–2000	9780, 11870, 15130, 15250, 17700

Vilnius (Lithuania) (54°30'N, 25°20'E)

Ann: "This is Radio Vilnius."

Stations:

```
7190 kHz,  120 kW          9735 kHz,  120 kW
9530 kHz,  120 kW          9745 kHz,  120 kW
9665 kHz,  120 kW
```

Broadcasts:

Time (UTC)	Freq (kHz)
2230–2300	(a) 7190, 9530, 9665, 9735, 9745

Legend:

(a) Saturday and Sunday only

VATICAN CITY

Vatican City (41°50′N, 12°28′E)

Ann: "Laudetur Jesus Christus. Praised be Jesus Christ. This is Vatican Radio."

Int: Bells of St. Paul's Cathedral; "Christus Vincit" played on a celesta.

Stations:

5995 kHz, 100 kW	11740 kHz, 100 kW
6190 kHz, 100 kW	11845 kHz, 100 kW
7235 kHz, 100 kW	15120 kHz, 100 kW
7250 kHz, 100 kW	15165 kHz, 100 kW
9605 kHz, 100 kW	17825 kHz, 100 kW
9615 kHz, 100 kW	17840 kHz, 100 kW
9645 kHz, 100 kW	17900 kHz, 100 kW
11700 kHz, 100 kW	21485 kHz, 100 kW
11705 kHz, 100 kW	

Broadcasts:

Time (UTC)	Freq (kHz)
0200–0215	(I) 5995, 9605, 11700
0200–0215	(II) 5995, 9605, 11700
0830–0930	(I) (a) 6190, 7250, 9645, 11740, 15120
1215–1230	(III) 17840, 21485
1300–1330	(III) 17900, 21485
1445–1500	(I) 6190, 7250, 9645, 11740
1615–1645	(IV) 11845, 15165, 17825
2130–2145	(I) 6190, 7250, 9645
2310–2330	(V) 7235, 9615, 11705

Legend:

(I) Beamed to Europe
(II) Beamed to North America
(III) Beamed to Africa
(VI) Beamed to Asia
(V) Beamed to Australia and New Zealand
(a) Mass

VIETNAM (Democratic Republic of)

Hanoi (21°30′N, 106°00′E)

Ann: "This is the voice of Vietnam from Hanoi, capital of the Democratic Republic of Vietnam."
Int: National anthem.

Stations:

10040 kHz, 50 kW	12035 kHz, 50 kW

Broadcasts:

Time (UTC)	Freq (kHz)
0100–0200	10040, 12035
1000–1100	10040, 12035
1300–1400	10040, 12035
1600–1700	10040, 12035
1800–1900	10040, 12035

YEMEN ARAB REPUBLIC

San'a (15°30′N, 44°30′E)

Ann: "Idha'at al Jumhuriyah al Arabiyah al Yamaniyah."

Stations:

4853 kHz, 5 kW
9780 kHz, 25 kW

Broadcasts:

Time (UTC)	Freq (kHz)
0300–0700	(a) 4853, 9780
1100–2200	(a) 4853, 9780

Legend:

(a) Arabic language only

YUGOSLAVIA

Belgrade (44°30′N, 20°09′E)

Ann: "This is Yugoslavia, Radio Belgrade calling."
Int: "Internationale."

Stations:

6100 kHz, 100 kW	11735 kHz, 100 kW
7240 kHz, 10 kW	15240 kHz, 100 kW
9620 kHz, 10 kW	

Broadcasts:

Time (UTC)	Freq (kHz)
1530–1600	9620, 11735, 15240
1830–1900	6100, 7240, 9620
2000–2030	6100, 7240, 9620
2200–2215	6100, 7240, 9620

ZAMBIA

Lusaka (15°30′S, 28°15′E)

Ann: "You are tuned to the general service of Radio Zambia."
Int: Music, "Call of the Fish Eagle."

Stations:

6060 kHz, 20 kW	9580 kHz, 50 kW
6165 kHz, 50 kW	11880 kHz, 50 kW

Broadcasts:

Time (UTC)	Freq (kHz)
1050–1215	9580, 11880
1550–2115	6060, 6165, 9580

3

Stations by Frequency

Freq	Location	Freq	Location
2425	Rhodesia, Salisbury	3973	Burundi, Bujumbura
3210	Mozambique, Lourenco Marques	3980	USA, Washington, DC
		3989	Great Britain, London
3265	Guyana (Rep), Georgetown	3990	USA, Washington, DC
		4740	Maldives, Male
3270	Dahomey, Cotonou	4775	Afghanistan, Kabul
3290	Guyana (Rep.), Georgetown	4777	Gabon, Libreville
		4780	Djibouti (People's Rep.), Djibouti
3300	Belize, Belize		
3315	Martinique, Fort-de-France	4785	Tanzania, Dar es Salaam
3316	Sierra Leone, Freetown	4795	Congo (People's Rep.), Brazzaville
3330	Gabon, Libreville		
3331	Comoro Islands, Moroni	4800	Lesotho, Maseru
3356	Botswana, Gaborone	4805	Kenya, Nairobi
3385	Guiana (French), Cayenne	4807	Sao Tome and Principe, Sao Tome
3396	Rhodesia, Salisbury	4820	Honduras (Rep.), Tegucigalpa
3425	Nepal, Kathmandu		
3560	Korea (North), Pyongyang	4845	Botswana, Gaborone
		4850	Mauritania, Nouakchott
3892	Korea (North), Pyongyang	4850	Mauritius, Port Louis
		4853	Yemen Arab Rep., San'a
3905	India, New Delhi	4865	Mozambique, Lourenco Marques
3915	Great Britain, London		
3945	New Hebrides, Vila	4870	Dahomey, Cotonou
3952	Great Britain, London	4875	South Africa, Johannesburg
3970	Cameroon, Buea		

Freq	Location	Freq	Location
4890	Papua New Guinea, Port Moresby	5980	South Africa, Johannesburg
4890	Senegal, Dakar	5990	Italy, Rome
4900	Burundi, Bujumbura	5990	Romania, Bucharest
4904	Chad, Ndjamena	5995	Australia, Melbourne
4907	Kampuchea, Phnom Penh	5995	USA, Washington, DC
		5995	Vatican City
4910	Guinea (Rep.), Conakry	5999	Great Britain, London
4940	Sri Lanka, Colombo	6000	Hungary, Budapest
4950	Sarawak, Kuching	6005	Australia, Melbourne
4970	Sabah, Kota Kinabalu	6005	Cameroon, Buea
4970	Sri Lanka, Colombo	6005	Great Britain, London
5010	Cameroon, Garoua	6005	Sri Lanka, Colombo
5010	Singapore, Singapore	6010	Germany (West), Cologne
5012	Rhodesia, Salisbury		
5015	Solomon Islands, Honiara	6010	USA, Washington, DC
		6015	Austria, Vienna
5035	Central African Republic, Bangui	6015	Norway, Oslo
		6015	USA, Washington, DC
5040	Burma, Rangoon	6020	Netherlands, Hilversum
5047	Togo, Lome	6020	USSR, Kiev
5070	Great Britain, London	6025	Portugal, Lisbon
5900	Israel, Jerusalem	6030	Uganda, Kampala
5900	USSR, Kiev	6030	USA, Washington, DC
5920	USSR, Kiev	6035	Australia, Melbourne
5930	Czechoslovakia, Prague	6035	USA, Washington, DC
5945	Albania, Tirana	6040	Canada, Montreal
5950	Guyana (Rep.), Georgetown	6040	Germany (West), Cologne
5955	Netherlands, Hilversum	6040	USA, Washington, DC
5955	United Nations	6045	Australia, Melbourne
5955	USA, Washington, DC	6045	Netherlands, Hilversum
5960	Germany (West), Cologne	6045	Panama, David
		6045	USA, Washington, DC
5960	Great Britain, London	6050	Great Britain, London
5960	Mongolian People's Rep., Ulan Bator	6050	Italy, Rome
		6055	Czechoslovakia, Prague
5965	Botswana, Gaborone	6055	United Nations
5965	Hungary, Budapest	6060	USA, Washington, DC
5965	Switzerland, Bern	6060	Zambia, Lusaka
5970	Canada, Montreal	6065	Germany (West), Cologne
5975	Great Britain, London		
5980	El Salvador, San Salvador	6065	Spain, Madrid
		6065	Sweden, Stockholm
5980	Guyana (Rep.), Georgetown	6070	Bulgaria, Sofia
		6070	USSR, Moscow

Freq	Location	Freq	Location
6075	Germany (West), Cologne	6155	Togo, Lome
6080	Belgium, Brussels	6155	USA, Oakland, Calif.
6080	Germany (East), Berlin	6165	Netherlands, Hilversum
6080	USA, Washington, DC	6165	Zambia, Lusaka
6085	Canada, Montreal	6170	Guiana (French), Cayenne
6085	India, New Delhi	6170	USSR, Kiev
6090	Kampuchea, Phnom Penh	6175	Great Britain, London
6090	Luxembourg, Luxembourg	6180	Great Britain, London
6090	United Nations	6180	Norway, Oslo
6095	Ecuador, Quito	6185	Ethiopia, Addis Ababa
6095	USA, Washington, DC	6185	Germany (West), Cologne
6100	Germany (West), Cologne	6185	USA, Washington, DC
6100	Spain, Madrid	6190	Romania, Bucharest
6100	Yugoslavia, Belgrade	6190	Vatican City
6105	Finland, Helsinki	6195	Canada, Montreal
6105	New Zealand, Wellington	6195	Great Britain, London
6105	Tanzania, Dar es Salaam	6200	Albania, Tirana
6110	USA, Washington, DC	6250	Equatorial Guinea, Malabo
6115	Germany (East), Berlin	6290	China, Peking
6115	Mozambique, Lourenco Marques	6338	Korea (North), Pyongyang
6115	Pakistan, Karachi	6383	Mongolian People's Rep., Ulan Bator
6120	Finland, Helsinki	6576	Korea (North), Pyongyang
6120	Haiti, Cap Haitien	6860	China, Peking
6120	Sweden, Stockholm	7035	China, Peking
6125	Canada, Montreal	7065	Albania, Tirana
6130	Ghana, Accra	7070	Albania, Tirana
6130	Sri Lanka, Colombo	7085	Pakistan, Karachi
6130	USA, Washington, DC	7100	USSR, Moscow
6135	Canada, Montreal	7105	Great Britain, London
6135	Poland, Warsaw	7105	Monaco, Monte Carlo
6135	Switzerland, Bern	7105	Nepal, Kathmandu
6135	Tahiti, Papeete	7105	Syrian Arab Rep., Damascus
6140	Canada, Montreal	7110	USA, Washington, DC
6140	Great Britain, London	7115	Bulgaria, Sofia
6140	USA, Washington, DC	7120	Chad, Ndjamena
6150	Romania, Bucharest	7120	China, Peking
6150	USSR, Moscow	7120	Great Britain, London
6155	Austria, Vienna	7120	Kenya, Nairobi
6155	Guinea (Rep.), Conakry	7125	Guinea (Rep.), Conakry
6155	Poland, Warsaw		

Freq	Location	Freq	Location
7130	Germany (West), Cologne	7225	India, New Delhi
7130	USSR, Moscow	7225	Philippines, Manila
7135	Great Britain, London	7225	Romania, Bucharest
7140	Great Britain, London	7230	Great Britain, London
7145	India, New Delhi	7230	USA, Washington, DC
7150	Germany (West), Cologne	7235	Italy, Rome
7150	Great Britain, London	7235	Vatican City
7150	Korea (South), Seoul	7240	Australia, Melbourne
7150	USSR, Kiev	7240	Korea (South), Seoul
7155	Canada, Montreal	7240	Netherlands, Hilversum
7155	China, Peking	7240	Yugoslavia, Belgrade
7155	Hungary, Budapest	7245	Algeria, Algiers
7155	Jordan, Amman	7245	Czechoslovakia, Prague
7160	Germany (West), Cologne	7245	Mauritania, Nouakchott
7160	Sarawak, Kuching	7245	Monaco, Monte Carlo
7160	USSR, Moscow	7245	USSR, Kiev
7165	Ethiopia, Addis Ababa	7250	USSR, Moscow
7165	France, Paris	7250	Vatican City
7165	USA, Washington, DC	7260	Comoro Islands, Moroni
7170	USA, Washington, DC	7260	Germany (East), Berlin
7170	USSR, Moscow	7260	New Hebrides, Vila
7180	Ethiopia, Addis Ababa	7270	Gabon, Libreville
7180	Great Britain, London	7270	Great Britian, London
7185	Burma, Rangoon	7270	Poland, Warsaw
7185	Germany (East), Berlin	7270	South Africa, Johannesburg
7185	Great Britain, London	7270	USA, Washington, DC
7190	Dahomey, Cotonou	7275	Italy, Rome
7190	USSR, Vilnius	7275	Nigeria, Ibadan
7195	Romania, Bucharest	7280	USA, Washington, DC
7195	USA, Washington, DC	7285	France, Paris
7200	Hungary, Budapest	7285	Germany (West), Cologne
7200	USA, Washington, DC	7285	Poland, Warsaw
7200	USSR, Moscow	7290	Italy, Rome
7203	Korea (North), Pyongyang	7290	USSR, Moscow
7210	Germany (West), Cologne	7295	USA, Washington, DC
7210	Great Britain, London.	7300	Albania, Tirana
7215	Brunei, Bandar Seri Begawan	7300	Germany (East), Berlin
7215	India, New Delhi	7310	USSR, Moscow
7215	USSR, Kiev	7315	China, Peking
7225	Germany (West), Cologne	7325	Great Britain, London
		7325	USA, Washington, DC
		7345	Czechoslovakia, Prague
		7350	USSR, Moscow
		7390	USSR, Moscow
		7395	Israel, Jerusalem

Freq	Location	Freq	Location
7412	Israel, Jerusalem	9530	USSR, Moscow
7580	Korea (North),	9530	USSR, Vilnius
	Pyongyang	9535	Angola, Luanda
7590	China, Peking	9540	Australia, Melbourne
7620	China, Peking	9540	Czechoslovakia, Prague
8072	Ethiopia, Addis Ababa	9540	Great Britain, London
9009	Israel, Jerusalem	9540	New Zealand,
9022	Iran, Tehran		Wellington
9030	China, Peking	9540	Poland, Warsaw
9380	China, Peking	9540	Romania, Bucharest
9410	Great Britain, London	9545	Germany (West),
9420	Korea (North),		Cologne
	Pyongyang	9545	Ghana, Accra
9425	Israel, Jerusalem	9545	Solomon Islands,
9435	Israel, Jerusalem		Honiara
9450	USSR, Moscow	9545	Syrian Arab Rep.,
9460	China, Peking		Damascus
9460	Pakistan, Karachi	9545	USA, Washington, DC
9470	China, Peking	9550	Australia, Melbourne
9475	Egypt, Cairo	9550	Finland, Helsinki
9480	Albania, Tirana	9550	Norway, Oslo
9500	Albania, Tirana	9550	USSR, Moscow
9500	Germany (East), Berlin	9555	El Salvador, San
9505	Czechoslovakia, Prague		Salvador
9505	France, Paris	9555	Kuwait, Kuwait
9505	Japan, Tokyo	9555	USSR, Moscow
9505	Philippines, Manila	9560	Bulgaria, Sofia
9505	Spain, Madrid	9560	Canada, Montreal
9510	Great Britain, London	9560	Ecuador, Quito
9510	Korea (North),	9560	Jordan, Amman
	Pyongyang	9565	Chile, Santiago
9510	Romania, Bucharest	9565	Germany (West),
9510	Taiwan, Taipei		Cologne
9515	Australia, Melbourne	9565	USA, Washington, DC
9515	Turkey, Ankara	9570	Australia, Melbourne
9515	Uganda, Kampala	9570	Great Britain, London
9515	USA, Washington, DC	9570	Romania, Bucharest
9520	Australia, Melbourne	9574	Mongolian People's
9520	Papua New Guinea,		Rep., Ulan Bator
	Port Moresby	9575	Australia, Melbourne
9525	Cuba, Havana	9575	Italy, Rome
9525	India, New Delhi	9580	Australia, Melbourne
9525	Korea (South), Seoul	9580	Great Britain, London
9525	Monaco, Monte Carlo	9580	Korea (South), Seoul
9525	Poland, Warsaw	9580	Kuwait, Kuwait
9530	Romania, Bucharest	9580	Zambia, Lusaka
9530	USA, Washington, DC	9585	France, Paris

Freq	Location	Freq	Location
9585	Hungary, Budapest	9650	Germany (West),
9585	Japan, Tokyo		Cologne
9585	Somalia, Mogadiscio	9650	Guinea (Rep.), Conakry
9585	South Africa,	9650	United Nations
	Johannesburg	9650	USSR, Moscow
9590	Germany (West),	9655	Canada, Montreal
	Cologne	9655	Hungary, Budapest
9590	Great Britain, London	9655	Thailand, Bangkok
9590	India, New Delhi	9660	Netherlands, Hilversum
9590	Switzerland, Bern	9660	USSR, Moscow
9600	Australia, Melbourne	9665	Germany (East), Berlin
9600	Great Britain, London	9665	Korea (South), Seoul
9600	Korea (South), Seoul	9665	USSR, Vilnius
9600	Taiwan, Taipei	9670	Great Britain, London
9605	Czechoslovakia, Prague	9670	Japan, Tokyo
9605	Germany (West),	9670	USA, Washington, DC
	Cologne	9675	Japan, Tokyo
9605	Japan, Tokyo	9675	Korea (South), Seoul
9605	Norway, Oslo	9675	Poland, Warsaw
9605	Sarawak, Kuching	9680	USA, Washington, DC
9605	Sweden, Stockholm	9680	USSR, Moscow
9605	Vatican City	9685	Cuba, Havana
9610	Algeria, Algiers	9685	Romania, Bucharest
9610	Ethiopia, Addis Ababa	9685	Taiwan, Taipei
9610	Mauritania, Nouakchott	9690	Argentina, Buenos Aires
9610	Norway, Oslo	9690	Germany (West),
9610	USSR, Kiev		Cologne
9610	USSR, Moscow	9690	Romania, Bucharest
9615	Chad, Ndjamena	9695	France, Paris
9615	Germany (West),	9695	Sweden, Stockholm
	Cologne	9700	Bulgaria, Sofia
9615	Vatican City	9700	Germany (West),
9620	Yugoslavia, Belgrade		Cologne
9625	Canada, Montreal	9700	Japan, Tokyo
9630	Czechoslovakia, Prague	9700	USA, Washington, DC
9630	Netherlands, Hilversum	9700	USSR, Moscow
9630	Sweden, Stockholm	9705	Bulgaria, Sofia
9630	United Nations	9705	USA, Washington, DC
9640	Great Britain, London	9710	Indonesia, Djakarta
9640	Korea (South), Seoul	9710	Italy, Rome
9640	Monaco, Monte Carlo	9710	Mauritius, Port Louis
9640	USA, Washington, DC	9710	United Nations
9645	Norway, Oslo	9710	USSR, Moscow
9645	Vatican City	9715	Congo (People's Rep.),
9650	Canada, Montreal		Brazzaville
9650	Germany (East), Berlin	9715	Netherlands, Hilversum
		9720	Korea (South), Seoul

Freq	Location	Freq	Location
9720	USSR, Moscow	9820	Korea (North),
9725	Belgium, Brussels		Pyongyang
9725	Burma, Rangoon	9860	China, Peking
9725	Switzerland, Bern	9895	Netherlands, Hilversum
9730	Germany (East), Berlin	9912	India, New Delhi
9730	India, New Delhi	9915	Great Britain, London
9730	Uganda, Kampala	9940	China, Peking
9730	USA, Washington, DC	9977	Korea (North),
9735	Germany (West),		Pyongyang
	Cologne	10040	Vietnam, Hanoi
9735	USSR, Vilnius	11445	China, Peking
9740	Czechoslovakia, Prague	11535	Korea (North),
9740	Great Britain, London		Pyongyang
9740	Portugal, Lisbon	11620	India, New Delhi
9740	USA, Washington, DC	11650	China, Peking
9745	Belgium, Brussels	11655	Israel, Jerusalem
9745	USSR, Vilnius	11675	China, Peking
9750	Great Britain, London	11675	Pakistan, Karachi
9750	Tahiti, Papeete	11685	China, Peking
9750	Tanzania, Dar es	11690	USSR, Kiev
	Salaam	11690	USSR, Moscow
9755	India, New Delhi	11700	USSR, Moscow
9755	USA, Washington, DC	11700	Vatican City
9758	Iraq, Baghdad	11705	Australia, Melbourne
9760	Australia, Melbourne	11705	France, Paris
9760	Germany (West),	11705	Japan, Tokyo
	Cologne	11705	New Zealand,
9760	Great Britain, London		Wellington
9760	USA, Washington, DC	11705	Sweden, Stockholm
9760	USSR, Moscow	11705	Vatican City
9765	Germany (West),	11710	Argentina, Buenos Aires
	Cologne	11715	Switzerland, Bern
9765	Taiwan, Taipei	11715	USA, Washington, DC
9768	Korea (North),	11720	Canada, Montreal
	Pyongyang	11720	Germany (East), Berlin
9770	Austria, Vienna	11720	Switzerland, Bern
9770	Haiti, Cap Haitien	11720	USSR, Moscow
9770	New Zealand,	11725	Australia, Melbourne
	Wellington	11725	Cuba, Havana
9770	USA, Washington, DC	11725	India, New Delhi
9775	Albania, Tirana	11725	Philippines, Manila
9780	China, Peking	11730	Netherlands, Hilversum
9780	USSR, Moscow	11735	Yugoslavia, Belgrade
9780	Yemen Arab Rep., San'a	11740	Australia, Melbourne
9805	Egypt, Cairo	11740	India, New Delhi
9815	Israel, Jerusalem	11740	Netherlands, Hilversum
		11740	USA, Washington, DC

Freq	Location	Freq	Location
11740	USSR, Moscow	11840	USA, Washington, DC
11740	Vatican City	11845	Kuwait, Kuwait
11745	Ecuador, Quito	11845	Sweden, Stockholm
11750	Great Britain, London	11845	Vatican City
11750	USSR, Moscow	11850	Germany (West),
11755	Finland, Helsinki		Cologne
11760	Cuba, Havana	11850	Ghana, Accra
11760	Great Britain, London	11850	India, New Delhi
11760	USA, Washington, DC	11850	Norway, Oslo
11765	Bulgaria, Sofia	11850	USSR, Moscow
11765	Germany (West),	11855	Canada, Montreal
	Cologne	11855	Czechoslovakia, Prague
11770	Great Britain, London	11855	Saudi Arabia, Riyadh
11770	India, New Delhi	11856	Mongolian People's
11775	India, New Delhi		Rep., Ulan Bator
11775	Switzerland, Bern	11860	Great Britain, London
11780	New Zealand,	11860	Korea (South), Seoul
	Wellington	11860	Norway, Oslo
11785	Germany (West),	11860	Taiwan, Taipei
	Cologne	11865	Germany (West),
11790	Australia, Melbourne		Cologne
11790	Indonesia, Djakarta	11870	Seychelles, Victoria
11790	Romania, Bucharest	11870	Switzerland, Bern
11790	USA, Washington, DC	11870	USSR, Moscow
11800	Poland, Warsaw	11875	Angola, Luanda
11800	USSR, Moscow	11875	Japan, Tokyo
11805	USA, Washington, DC	11875	Philippines, Manila
11810	Australia, Melbourne	11880	India, New Delhi
11810	Germany (West),	11880	Spain, Madrid
	Cologne	11880	Turkey, Ankara
11810	India, New Delhi	11880	Zambia, Lusaka
11815	Japan, Tokyo	11885	Pakistan, Karachi
11815	Poland, Warsaw	11890	Bangladesh, Dacca
11820	Great Britain, London	11890	France, Paris
11820	USA, Washington, DC	11890	Germany (East), Berlin
11825	Canada, Montreal	11890	Philippines, Manila
11825	India, New Delhi	11895	Norway, Oslo
11825	Tahiti, Papeete	11895	Senegal, Dakar
11825	Taiwan, Taipei	11900	South Africa,
11825	United Nations		Johannesburg
11830	Romania, Bucharest	11905	Germany (West),
11830	USSR, Moscow		Cologne
11835	Haiti, Cap Haitien	11905	Sweden, Stockholm
11840	Australia, Melbourne	11905	Switzerland, Bern
11840	Germany (East), Berlin	11905	Thailand, Bangkok
11840	Poland, Warsaw	11910	Algeria, Algiers
11840	Romania, Bucharest	11910	Hungary, Budapest

Freq	Location	Freq	Location
11915	Ecuador, Quito	15110	Czechoslovakia, Prague
11920	Ivory Coast, Abidjan	15110	Finland, Helsinki
11920	Philippines, Manila	15110	Pakistan, Karachi
11925	Germany (West),	15115	Ecuador, Quito
	Cologne	15120	Nigeria, Ibadan
11925	USA, Washington, DC	15120	Poland, Warsaw
11930	Cuba, Havana	15120	Sweden, Stockholm
11930	France, Paris	15120	Vatican City
11930	United Nations	15125	Germany (East), Berlin
11935	Australia, Melbourne	15130	New Zealand,
11935	USA, Washington, DC		Wellington
11940	Bangladesh, Dacca	15130	USSR, Moscow
11940	Belgium, Brussels	15135	Norway, Oslo
11940	Canada, Montreal	15140	Australia, Melbourne
11940	Romania, Bucharest	15140	Switzerland, Bern
11940	Singapore, Singapore	15145	Germany (East), Berlin
11945	China, Peking	15150	USA, Washington, DC
11945	Liberia, Monrovia	15155	United Nations
11950	Japan, Tokyo	15160	Hungary, Budapest
11950	Liberia, Monrovia	15160	Seychelles, Victoria
11955	Great Britain, London	15165	India, New Delhi
11960	New Zealand,	15165	Vatican City
	Wellington	15170	Germany (East), Berlin
11960	USSR, Moscow	15170	Tahiti, Papeete
11965	Guinea (Rep.), Conakry	15175	Norway, Oslo
11970	Germany (West),	15180	Australia, Melbourne
	Cologne	15180	Bangladesh, Dacca
11985	Albania, Tirana	15180	USSR, Moscow
11990	Czechoslovakia, Prague	15185	Netherlands, Hilversum
12000	USSR, Moscow	15185	Nigeria, Ibadan
12010	China, Peking	15185	USA, Red Lion, Pa.
12020	USSR, Moscow	15185	USA, Washington, DC
12035	Vietnam, Hanoi	15190	Congo (People's Rep.),
12040	Great Britain, London		Brazzaville
12045	Israel, Jerusalem	15190	India, New Delhi
12050	USSR, Moscow	15195	Afghanistan, Kabul
12055	China, Peking	15195	Japan, Tokyo
12055	Israel, Jerusalem	15205	India, New Delhi
12095	Great Britain, London	15205	USA, Washington, DC
15060	China, Peking	15210	USSR, Moscow
15070	Great Britain, London	15220	South Africa,
15080	India, New Delhi		Johannesburg
15100	Israel, Jerusalem	15220	USA, Washington, DC
15100	USSR, Kiev	15225	Hungary, Budapest
15100	USSR, Moscow	15225	Taiwan, Taipei
15105	Great Britain, London	15235	Japan, Tokyo
15105	Japan, Tokyo	15240	Australia, Melbourne

Freq	Location	Freq	Location
15240	Sweden, Stockholm	15365	Romania, Bucharest
15240	Yugoslavia, Belgrade	15380	Romania, Bucharest
15250	Germany (East), Berlin	15380	USSR, Moscow
15250	Romania, Bucharest	15390	Philippines, Manila
15250	USA, Washington, DC	15395	Czechoslovakia, Prague
15250	USSR, Moscow	15400	Bangladesh, Dacca
15260	Great Britain, London	15410	Australia, Melbourne
15265	Finland, Helsinki	15410	Germany (West),
15270	Bangladesh, Dacca		Cologne
15270	China, Peking	15410	Great Britain, London
15270	Japan, Tokyo	15410	United Nations
15275	Germany (West),	15415	Hungary, Budapest
	Cologne	15415	United Nations
15275	Malaysia, Kuala Lumpur	15420	Algeria, Algiers
15275	Poland, Warsaw	15420	Great Britain, London
15280	Great Britain, London	15420	Japan, Tokyo
15280	Philippines, Manila	15420	USSR, Moscow
15285	Ghana, Accra	15425	France, Paris
15285	United Nations	15425	Taiwan, Taipei
15290	Australia, Melbourne	15430	Japan, Tokyo
15290	USA, Washington, DC	15430	Switzerland, Bern
15300	Cuba, Havana	15430	USA, Washington, DC
15300	Japan, Tokyo	15435	Great Britain, London
15300	USSR, Moscow	15435	Tanzania, Dar es
15305	Sweden, Stockholm		Salaam
15305	Switzerland, Bern	15440	Philippines, Manila
15305	United Nations	15465	Israel, Jerusalem
15305	USA, Red Lion, Pa.	15485	Israel, Jerusalem
15310	Ecuador, Quito	16085	Ethiopia, Addis Ababa
15310	Great Britain, London	17387	India, New Delhi
15310	Japan, Tokyo	17665	Pakistan, Karachi
15320	Australia, Melbourne	17695	Great Britain, London
15320	Germany (East), Berlin	17700	Germany (East), Berlin
15325	Canada, Montreal	17700	Netherlands, Hilversum
15325	Japan, Tokyo	17700	USSR, Moscow
15325	Pakistan, Karachi	17705	Great Britain, London
15325	Uganda, Kampala	17710	Hungary, Budapest
15330	USA, Washington, DC	17710	USA, Washington, DC
15335	India, New Delhi	17715	Germany (West),
15340	Portugal, Lisbon		Cologne
15340	Romania, Bucharest	17715	Hungary, Budapest
15345	Kuwait, Kuwait	17720	Hungary, Budapest
15345	Norway, Oslo	17720	Taiwan, Taipei
15345	Romania, Bucharest	17725	Japan, Tokyo
15345	Taiwan, Taipei	17730	Malagasy Republic,
15355	Australia, Melbourne		Tananarive
15365	Great Britain, London	17730	Sweden, Stockholm

Freq	Location	Freq	Location
17750	United Nations	17880	Portugal, Lisbon
17765	Germany (West),	17885	Cuba, Havana
	Cologne	17885	Great Britain, London
17765	USA, Washington, DC	17890	Bangladesh, Dacca
17780	Ecuador, Quito	17890	Taiwan, Taipei
17780	Germany (West),	17895	Portugal, Lisbon
	Cologne	17895	USA, Washington, DC
17780	Hungary, Budapest	17900	Vatican City
17780	South Africa,	17920	Egypt, Cairo
	Johannesburg	17935	Portugal, Lisbon
17785	Hungary, Budapest	21465	Germany (East), Berlin
17785	USA, Washington, DC	21470	Great Britain, London
17790	Great Britain, London	21485	USA, Washington, DC
17790	USA, Washington, DC	21485	Vatican City
17795	Australia, Melbourne	21500	Germany (West),
17795	Sweden, Stockholm		Cologne
17800	Germany (East), Berlin	21510	USSR, Moscow
17810	Netherlands, Hilversum	21515	Philippines, Manila
17810	Philippines, Manila	21520	Switzerland, Bern
17815	Israel, Jerusalem	21525	Hungary, Budapest
17820	Canada, Montreal	21535	South Africa,
17820	USA, Washington, DC		Johannesburg
17825	Bulgaria, Sofia	21540	Germany (East), Berlin
17825	Germany (West),	21540	Germany (West),
	Cologne		Cologne
17825	Japan, Tokyo	21545	Ghana, Accra
17825	Vatican City	21570	Australia, Melbourne
17830	Pakistan, Karachi	21590	Pakistan, Karachi
17830	Switzerland, Bern	21600	Germany (West),
17840	Czechoslovakia, Prague		Cologne
17840	Great Britain, London	21610	Great Britain, London
17840	Norway, Oslo	21610	USA, Washington, DC
17840	Switzerland, Bern	21660	Great Britain, London
17840	Vatican City	21670	United Nations
17865	United Nations	21670	USA, Washington, DC
17870	Australia, Melbourne	21685	Bangladesh, Dacca
17870	USA, Washington, DC	21690	Sweden, Stockholm
17875	Germany (West),	21700	Czechoslovakia, Prague
	Cologne	21710	Great Britain, London
17875	USA, Red Lion, Pa.	21720	Ghana, Accra
17880	Great Britain, London	21730	Norway, Oslo
17880	Japan, Tokyo		

4

International Shortwave
News Broadcasts

International shortwave radio broadcasting provides a means by which the people of other nations may learn the history, culture, and folklore of the originating country. But more important, it is a dynamic medium by which national and international news and commentaries can be broadcast to the ears of millions of listeners outside the borders of a country.

The purpose of this section is to provide the enthusiast with a convenient listing of all English-language news broadcasts transmitted from the capitals of the world. When, in some instances, a specific country does not broadcast in English, the country and the language used for the news broadcast are listed in Section 2.

To use this section, note the time a specific country broadcasts the news, then refer to Section 2 for further details of the transmission. All time indications are in UTC.

0000

Albania
Australia
Bulgaria
China
Great Britain
Guyana (Rep.)
Haiti
Japan
Luxembourg
New Zealand
Norway
Pakistan
Sabah
Sarawak
Thailand
USA (Washington, AFRTS)
USA (Washington, VOA)
USSR (Moscow)

0015

Australia
Belgium

0030

Philippines (FEBC)
Singapore
Sweden
USA (Washington, AFRTS)
USSR (Kiev)

0045

India

0100

Belize
Canada
Chile
China
Cuba
Czechoslovakia
Ecuador
Germany (East)
Italy
Japan
Luxembourg
Monaco
Philippines (VOP)
Spain
Taiwan
USA (Oakland, WYFR)
USA (Washington, AFRTS)
USA (Washington, VOA)
USSR (Moscow)
Vietnam

0115

Sri Lanka

0130

Albania
Australia
Germany (West)
Japan
Philippines (FEBC)
Romania
Solomon Islands
USA (Washington, AFRTS)

0145

Switzerland

0200

Australia
Belize
Burma
China
Cuba
Ecuador
Egypt
Great Britain
Guyana (Rep.)
Hungary
Japan
Luxembourg
Netherlands
Norway
Poland
Spain
Sri Lanka
USA (Oakland, WYFR)
USA (Washington, AFRTS)
USA (Washington, VOA)
USSR (Moscow)
Vatican City

0210

Chile

0230

Albania
Australia
Germany (East)
Pakistan
Papua New Guinea
Philippines (FEBC)
Portugal
Sweden
United Nations
USA (Washington, AFRTS)

0300

Argentina
Australia
Belize
China
Cuba
Czechoslovakia
Egypt
Finland

Great Britain
Honduras
Hungary
Japan
Korea (South)
Netherlands
South Africa
Spain
Taiwan
USA (Oakland, WYFR)
USA (Washington, AFRTS)
USA (Washington, VOA)
USSR (Kiev)
USSR (Moscow)

0310

Chile

0330

Albania
Austria
Cuba
Finland
Germany (East)
Papua New Guinea
Philippines (FEBC)
USA (Washington, AFRTS)
USSR (Moscow)

0340

Pakistan

0350

Italy

0355

Spain

0400

Australia
Belize
Bulgaria
Canada
Cuba
Ecuador
Great Britain
Honduras
Hungary
Japan

Kenya
Mauritius
Norway
Rhodesia
Romania
South Africa
Tanzania
USA (Washington, AFRTS)
USA (Washington, VOA)
Yemen Arab Rep.

0415

Thailand

0420

South Africa

0425

Italy
Thailand

0430

Albania
Brunei
Chad
Germany (West)
Pakistan
Philippines (FEBC)
Switzerland
United Nations
USA (Washington, AFRTS)
USSR (Moscow)

0445

Australia
Bangladesh
Germany (East)
USSR (Moscow)

0500

Australia
Belize
Burundi
Canada
Cuba
Ecuador
Gabon
Great Britain
Honduras

Israel
Japan
Korea (North)
Lesotho
Mozambique
Netherlands
Portugal
Sabah
Tanzania
USA (Washington, AFRTS)
USA (Washington, VOA)

0510

Botswana

0515

USA (Washington, AFRTS)
USSR (Moscow)

0530

Australia
Germany (West)
Kuwait
Papua New Guinea
Romania
USA (Washington, AFRTS)
USA (Washington, VOA)
USSR (Moscow)

0545

Cameroon (Buea)
United Nations

0600

Argentina
Australia
Botswana
Gabon
Germany (West)
Great Britain
Japan
Kenya
New Hebrides
New Zealand
Norway
Rhodesia
Sao Tome and Principe
Seychelles
Singapore

USA (Washington, AFRTS)
USA (Washington, VOA)

0609

Liberia

0615

Dahomey
South Africa
Syrian Arab Rep.

0620

Canada

0630

Albania
Algeria
Cuba
Gabon
Korea (South)
Malaysia
Netherlands
Poland
Sierra Leone
United Nations
USA (Washington, AFRTS)
USA (Washington, VOA)
USSR (Moscow)

0645

Cameroon (Buea)
Germany (East)
Romania

0700

Albania
Australia
Burma
Burundi
Canada
Cuba
Dahomey
Ghana
Great Britain
Japan
Korea (North)
Kuwait
Nigeria
Rhodesia

Sao Tome and Principe
Seychelles
Sierra Leone
Solomon Islands
Switzerland
Syrian Arab Rep.
United Nations
USA (Washington, AFRTS)
USA (Washington, VOA)
Zambia

0707

United Nations

0709

Liberia

0730

Algeria
Australia
Czechoslovakia
Mauritania
Monaco
Nigeria
USA (Washington, AFRTS)

0740

Canada

0745

Sri Lanka
United Nations

0800

Australia
Dahomey
Ghana
Great Britain
Japan
Netherlands
New Hebrides
Norway
Papua New Guinea
Rhodesia
USA (Washington, AFRTS)
USSR (Moscow)

0830

Australia
China

Czechoslovakia
Malaysia
New Zealand
Philippines (FEBC)
USA (Washington, AFRTS)
Vatican City

0845

Australia
United Nations

0900

Algeria
Australia
Great Britain
Indonesia
Japan
Kenya
Monaco
Papua New Guinea
Philippines (VOP)
Rhodesia
Switzerland
USA (Washington, AFRTS)
USSR (Moscow)

0905

Botswana

0930

Albania
Comoro Islands
Finland
Germany (West)
Japan
Netherlands
Philippines (FEBC)
Singapore
USA (Washington, AFRTS)

1000

Algeria
Australia
Canada
Gabon
India
Japan
Jordan
Kenya

Korea (North)
Korea (South)
Martinique
New Zealand
Sao Tome and Principe
Saudi Arabia
Tanzania
United Nations
USA (Washington, AFRTS)
USSR (Moscow)
Vietnam

1010

Rhodesia

1030

Guiana (French)
Hungary
Monaco
Singapore
Thailand
USA (Washington, AFRTS)

1045

Germany (West)

1050

Chile

1055

Thailand

1100

Albania
Algeria
Australia
Djibouti
Great Britain
Haiti
Indonesia
Jordan
Korea (South)
Mauritania
Pakistan
Papua New Guinea
Sabah
Sarawak
Solomon Islands
Somalia

South Africa
Sweden
Switzerland
USA (Washington, AFRTS)
USA (Washington, VOA)
USSR (Moscow)

1115

Canada
Japan
Jordan
Rhodesia
Thailand
Zambia

1125

Botswana

1130

Afghanistan
Angola
Korea (South)
Lesotho
Singapore
USA (Washington, AFRTS)

1135

Botswana

1145

Canada

1200

Algeria
Australia
China
Dahomey
Gabon
Germany (East)
Germany (West)
Haiti
Hungary
Israel
Japan
Jordan
Kenya
Korea (North)
Korea (South)
Liberia

Martinique
Norway
Poland
Romania
Sabah
Sao Tome and Principe
Saudi Arabia
USA (Washington, AFRTS)
USA (Washington, VOA)
Vatican City
Yemen Arab Rep.

1210

Australia
Chile

1215

Vatican City

1220

Mongolian People's Rep.

1230

Australia
Austria
Bangladesh
Chad
Ecuador
Guinea (Rep.)
Poland
Sweden
USA (Washington, AFRTS)
USSR (Moscow)

1240

Dahomey

1245

Belize
Sri Lanka
Togo
Vatican City

1300

Algeria
Australia
Belize
Ecuador
Germany (West)

Great Britain
Guyana (Rep.)
Japan
Kenya
Rhodesia
Romania
USA (Oakland, WYFR)
USA (Washington, AFRTS)
USA (Washington, VOA)
Vatican City
Vietnam

1315

Brunei
Egypt
Germany (East)
Switzerland

1330

Chile
India
Philippines (FEBC)
Singapore
USA (Washington, AFRTS)
USSR (Moscow)

1400

Afghanistan
Albania
Australia
Belize
Canada
China
Ecuador
Germany (East)
Ghana
Great Britain
Guyana (Rep.)
Japan
Kenya
Korea (South)
Mauritania
Netherlands
Norway
Philippines (VOP)
Portugal
Rhodesia
Sao Tome and Principe
Sweden

USA (Washington, AFRTS)
USA (Washington, VOA)

1430

Czechoslovakia
Finland
Guinea (Rep.)
Hungary
Philippines (FEBC)
Uganda
United Nations
USA (Washington, AFRTS)

1435

Angola
Nepal

1445

Burma
Cameroon (Buea)
Ghana
South Africa
Vatican City

1500

Algeria
Australia
Belize
Ecuador
Gabon
Great Britain
Guyana (Rep.)
Japan
Jordan
Kenya
Malagasy Republic
Maldives
Romania
Sabah
Sao Tome and Principe
Sarawak
USA (Washington, AFRTS)
USA (Washington, VOA)
USSR (Moscow)

1505

Monaco

1530

Albania
Czechoslovakia
Germany (East)
Lesotho
Nigeria
Philippines (FEBC)
Seychelles
Singapore
Switzerland
USA (Washington, AFRTS)
Yugoslavia

1545

Djibouti
Sri Lanka

1550

Vatican City

1600

Algeria
Australia
China
Finland
Great Britain
Japan
Jordan
Kenya
Korea (South)
Liberia
Maldives
Norway
Poland
Portugal
Rhodesia
Sao Tome and Principe
South Africa
Sweden
Tanzania
USA (Washington, AFRTS)
USA (Washington, VOA)
Vietnam
Zambia

1609

Liberia

1615

Seychelles
Uganda
Vatican City

1630

Albania
Burundi
Comoro Islands
Czechoslovakia
Guiana (French)
Japan
Nigeria
Pakistan
Poland
USA (Washington, AFRTS)

1645

Ghana

1700

Algeria
Australia
Belize
Finland
France
Great Britain
Guyana (Rep.)
Japan
Jordan
Kuwait
Maldives
Martinique
Portugal
Sri Lanka
Taiwan
USA (Oakland, WYFR)
USA (Red Lion, WINB)
USA (Washington, AFRTS)
USA (Washington, VOA)
Yemen Arab Rep.

1709

Liberia

1715

Belgium
Mongolian People's Rep.

1720

Germany (West)

1725

Guinea (Rep.)
Liberia

1730

Albania
Cameroon (Buea)
Kuwait
Romania
USA (Washington, AFRTS)

1745

India
Pakistan
Rhodesia
Sao Tome and Principe

1800

Australia
Canada
Finland
Gabon
Germany (East)
Great Britain
Guyana (Rep.)
Japan
Kenya
Korea (North)
Korea (South)
Liberia
Mauritania
Mauritius
Mozambique
New Zealand
Norway
Portugal
Sabah
Tanzania
Uganda
United Nations
USA (Washington, AFRTS)
USA (Washington, VOA)
Vietnam
Zambia

1810

Botswana

1815

Bangladesh
Germany (East)
Ghana

1830

Albania
Belize
Cameroon (Garoua)
Ivory Coast
Japan
Netherlands
New Zealand
Nigeria
Poland
Portugal
Sweden
United Nations
USA (Washington, AFRTS)
USSR (Moscow)
Yugoslavia

1840

Senegal

1900

Algeria
Australia
Bulgaria
Chad
Czechoslovakia
Dahomey
Equatorial Guinea
Guyana (Rep.)
Ivory Coast
Japan
Korea (North)
Liberia
New Zealand
Nigeria
Rhodesia
Saudi Arabia
Tahiti
United Nations
USA (Washington, AFRTS)
USA (Washington, VOA)

1905

Botswana

1920

Dahomey
India

1925

Guinea (Rep.)

1930

Albania
Bulgaria
China
Finland
Germany (East)
Germany (West)
Iraq
Mauritania
Romania
Sao Tome and Principe
USA (Red Lion, WINB)
USA (Washington, AFRTS)
USSR (Kiev)
USSR (Moscow)

1945

India
Yemen Arab Rep.

1950

Togo

2000

Algeria
Australia
Belize
Czechoslovakia
Equatorial Guinea
Finland
Germany (East)
Ghana
Great Britain
Guinea (Rep.)
Guyana (Rep.)
Iran
Israel
Japan

Kenya
Korea (North)
Korea (South)
Martinique
Netherlands
Norway
Poland
Rhodesia
Sabah
Saudi Arabia
USA (Washington, AFRTS)
USA (Washington, VOA)
USSR (Moscow)
Yugoslavia
Zambia

2010

Cuba

2015

Japan

2020

India

2025

Italy

2030

Albania
Cameroon (Buea)
Central African Rep.
China
Equatorial Guinea
Poland
Syrian Arab Rep.
Uganda
USA (Washington, AFRTS)
USSR (Moscow)

2045

Ghana
India
Liberia
Sweden

2050

Cuba

2100

Australia
Belize
Bulgaria
Canada
Congo (People's Rep.)
Cuba
Dahomey
Gabon
Germany (West)
Ghana
Guyana (Rep.)
Japan
Korea (South)
Pakistan
Rhodesia
Romania
Saudi Arabia
South Africa
Sweden
Switzerland
Syrian Arab Rep.
USA (Oakland, WYFR)
USA (Washington, AFRTS)
USA (Washington, VOA)
USSR (Moscow)
Zambia

2115

Chad
Congo (People's Rep.)
Germany (East)

2130

Bulgaria
China
Czechoslovakia
Hungary
Netherlands
Spain
Taiwan
USA (Washington, AFRTS)
Vatican City

2145

Australia
Egypt
South Africa

2150

Congo (People's Rep.)

2200

Albania
Algeria
Australia
Canada
Equatorial Guinea
Gabon
Great Britain
Guinea (Rep.)
Guyana (Rep.)
Italy
Japan
New Zealand
Norway
Sao Tome and Principe
Solomon Islands
Turkey
United Nations
USA (Washington, AFRTS)
USA (Washington, VOA)
USSR (Moscow)
Yugoslavia

2210

Mauritania

2215

Vatican City

2230

Guiana (French)
Israel
Poland
South Africa
USA (Washington, AFRTS)
USSR (Vilnius)

2235

Congo (People's Rep.)

2245

Australia
India
Mauritania
New Zealand
Papua New Guinea

2250

Chile

2300

Algeria
Argentina
Australia
Belize
Brunei
Great Britain
Guyana (Rep.)
Japan
Korea (North)
Korea (South)
Martinique
USA (Washington, AFRTS)
USA (Washington, VOA)
USSR (Moscow)

2310

Brunei
Vatican City

2315

South Africa

2320

India
Vatican City

2330

Australia
Guiana (French)
Guinea (Rep.)
Indonesia
New Zealand
Philippines (FEBC)
Sarawak
Singapore
Solomon Islands
Thailand
USA (Washington, AFRTS)

2345

Japan

2355

Algeria
Guinea (Rep.)

5

Verification Reports
[QSL-ing]

Most international shortwave broadcasting stations are interested in knowing how well their transmissions are being received by listeners in other countries, particularly in those areas to which the signals are beamed. Listener reports are useful to the station engineering staff in selecting frequencies to avoid interference or atmospheric disturbances. They are also useful to the program manager in determining listener reactions to the program. In return for these reception reports, a station will usually mail to the sender a verification card (QSL card).

To make your report as brief but as useful as possible, the following items must be included (preferably typewritten).

1. Date and time (UTC) the transmission was heard and the length of time you listened, such as "0700–0800 hours UTC, July 10, 1978."
2. Reception report using SINPO code (Chart 1, page 131).
3. Make, type, and model of receiver used.
4. Type of antenna used (height, length, whether it is erected in the clear, etc.).
5. A brief description of what the program contained, such as the musical selections played, subject material of the commentary, or any other program data. The station staff uses this information to identify the program as positively being theirs and to verify that you did, indeed, hear it.
6. Comments regarding reception (unusual interference, fading characteristics, modulation quality, and any other transmission irregularities).

7. Program comments and suggestions.
8. Your name and complete mailing address.

Allow seven to ten weeks for receipt of your QSL card. It is advisable to send your report via air mail. Consult your local postal authorities for postal rates. Mail your report to the address given in the following list.

AFGHANISTAN

Radio Afghanistan
P.O. Box 544
Kabul, Afghanistan

ALBANIA

Radio & Television Albania
Rue Ismail Quemal
Tirana, Albania

ALGERIA

Radio Algeria
21, Boulevard des Martyrs
Algiers, Algeria

ANGOLA

Director of Telegraph &
Telephone
Luanda, Angola

ARGENTINA

Radio R.A.E.
Sarmiento 151
Buenos Aires, Argentina

AUSTRALIA

Radio Australia
P. O. 428 G
G.P.O.
Melbourne 3001, Australia

AUSTRIA

Radio Austria (ORF)
Shortwave Service
A-1136
Vienna, Austria

BANGLADESH

Radio Bangladesh
External Service
145/A, Rd. No. 2

Dhanmondi R/A
Dacca 2, People's Republic
of Bangladesh

BELGIUM

Belgium Radio & Television
Service
Station ORU
P.O. Box 26, B-1000
Brussels, Belgium

BELIZE

Radio Belize
P.O. Box 89
Belize City
Belize

BOTSWANA

Radio Botswana
P.O. Box 52
Gaborone, Botswana

BRUNEI

Radio Brunei
Dept. of Broadcasting &
Information
Bandar Seri Begawan, Brunei

BULGARIA

Radio Sofia
Committee of Radio &
Television
Culture & Art
4 Bd. Dragan Tsankov
Sofia, Bulgaria

BURMA

Burma Broadcasting Service
Prome Rd. Kamayut P.O.
Rangoon, Burma

BURUNDI

Radio CORDAC
B.P. 1140 Bujumbura
Burundi

CAMEROON

Radio Buea
P.O. 86
Buea, Cameroon

Radio Garoua
B.P. 103
Garoua, Cameroon

CANADA

Radio Canada International
P.O. Box 6000
Montreal, Canada H3C 3A8

CENTRAL AFRICAN REPUBLIC

National Central African Radio
B.P. 940
Bangui, Centra' African
    Republic

CHAD

National Tchadienne Radio
B.P. 892
Ndjamena, Republic of Chad

CHILE

Radio Nacional del Chile
Casilla 244V
Santiago, Chile

CHINA (People's Republic of; Mainland)

Radio Peking
Broadcasting Administration
Fu Hsin Men, Peking
People's Rep. of China

COMORO ISLANDS

Comoro International Radio
B.P. 250
Moroni
Comoro Islands

CONGO (People's Rep. of the)

Congo Radio & Television
B.P. 2241
Brazzaville, People's Rep. of
    the Congo

CUBA

Radio Havana
P.O. Box 7062
Havana, Cuba

CZECHOSLOVAKIA

Radio Prague
Czechoslovakia Radio
Prague 2, Czechoslovakia

DAHOMEY

Radiodiff. du Dahomey
B.P. 366
Cotonou, Republic of Dahomey

DJIBOUTI

Director of Radio and
    Television
B.P. 97
Djibouti, People's Republic of
    Djibouti

ECUADOR

The Voice of the Andes, HCJB
Casilla 691
Quito, Ecuador

EGYPT

Radio Cairo
Radio & TV Building
P.O. Box 1186, Kornish Road
Cairo, Egypt

EL SALVADOR

National Radio of El Salvador
2a Avenida Sur 113
San Salvador, El Salvador

EQUATORIAL GUINEA

Radio Malabo
Apt. 195
Malabo, Equatorial Guinea

ETHIOPIA

Voice of Revolutionary
Ethiopia
Addis Ababa, Ethiopia

FINLAND

Finnish Broadcasting Co.
Kesakatu 2, 00260
Helsinki 26, Finland

FRANCE

Radio France International
B.P. 9516, F-75016
Paris, France

GABON

Radiodiff. Television
Gabonaise
La Voix de la Renovation
P.B. 10150
Libreville, Gabon

GERMANY (Democratic Republic; East)

Radio Berlin International
116 Berlin, Nalepastrasse
18-50
German Democratic Republic

GERMANY (Federal Republic; West)

Deutsche Welle
The Voice of Germany
P.O. Box 10 04 44
5 Cologne 1
Federal Republic of Germany

GHANA

Ghana Broadcasting Corp.
P.O. Box 1633
Accra, Ghana

GREAT BRITAIN

BBC London
British Broadcasting Corp.
Broadcasting House
London W1A 1AA,
Great Britain

GUIANA (French)

France Region 3
Office of Radiodiff. and
Television
B.P. 336
Cayenne, French Guiana

GUINEA (Republic of)

La Voix de la Revolution
Radiodiff. Nationale
B.P. 617
Conakry, Republic of Guinea

GUYANA (Republic)

Action Radio, GBS
Guyana Broadcasting Service
P.O. Box 560
Georgetown, Republic of
Guyana

Radio Demerara
Guyana Broadcasting Co., Ltd.
P.O. Box 561
Georgetown, Republic of
Guyana

HAITI

Radio Station 4VEH
Box 1
Cap Haitien, Haiti

HONDURAS (Republic of)

Baptist Home Mission Society
Radio
Ap. 145-C
Tegucigalpa, Republic of
Honduras

HUNGARY

Radio Budapest
Brody Sandor 5-7
H–1800
Budapest, Hungary

INDIA

All India Radio
P.O. Box 500
New Delhi, India

INDONESIA

The Voice of Indonesia
P.O. Box 157
Djakarta, Indonesia

IRAN

Radio Iran
External Broadcasting
P.O. Box 33-200
Tehran, Iran

IRAQ

Radio Baghdad
Iraqi Broadcasting & TV
Establishment
Salihiya
Baghdad, Iraq

ISRAEL

Israel Broadcasting Authority
Overseas Service
P.O. Box 1082
Jerusalem, Israel

ITALY

Italian Radio & Television
Service
Viale Mazzini 14,
00195 Rome, Italy

IVORY COAST

Ivory Coast Broadcasting
System
Radio Abidjan
B.P. 2261
Abidjan, Ivory Coast

JAPAN

Radio Japan, NHK
2-2-1 Jinnan
Shibuya-ku
Tokyo, Japan

JORDAN

Radio Jordan
Hashemite Kingdom
Broadcasting Service
P.O. Box 909
Amman, Jordan

KAMPUCHEA (Cambodia)

Voice of the National United
Front of Kampuchea
28, Ave. Sandech Choun Nath
Phnom-Penh, Democratic
Kampuchea

KENYA

Voice of Kenya
Box 30456
Nairobi, Kenya

KOREA (Democratic People's Republic; North)

Radio Pyongyang
Korean Central Broadcasting
Committee
Pyongyang, Democratic
People's Republic of Korea

KOREA (Republic; South)

Radio Korea
Korean Broadcasting Corp.
8, Yejang-Dong
Joong-gu
Seoul, Republic of Korea

KUWAIT

Radio Kuwait
P.O. Box 397
Kuwait

LESOTHO

Radio Lesotho
P.O. Box 552
Maseru, Lesotho

LIBERIA

Radio Station ELWA
Box 192
Monrovia, Liberia

LUXEMBOURG

Radio Luxembourg
Compagnie Luxembourgeoise
de Telediffusion
Villa Louvigny
Luxembourg

MALAGASY REPUBLIC

Radio-Television Malagasy
B.P. 442
Tananarive, Malagasy

MALAYSIA

Radio Malaysia
Department of Broadcasting
Angkasapuri
Kuala Lumpur, Malaysia

MALDIVES

Radio Maldives
Department of Information &
Broadcasting
Male, Republic of the
Maldives

MARTINIQUE

France Region 3
Office of Radiodiff. and
Television
B.P. 662
Fort-de-France, Martinique

MAURITANIA

National Radio of Mauritania
B.P. 200
Nouakchott, Republic of
Mauritania

MAURITIUS

Mauritius Broadcasting
Corporation
Broadcasting House
Port Louis, Mauritius

MONACO

Trans World Radio
P.O. Box 141
Monte Carlo, Monaco

MONGOLIAN PEOPLE'S REPUBLIC

Ulan Bator Radio
CPO Box 365
Ulan Bator
People's Republic of Mongolia

MOZAMBIQUE

Radio Club of Mozambique
P.O. Box 594
Lourenco Marques,
Mozambique

NEPAL

Radio Nepal
P.O. Box 634, Singha Durbar
Department of Broadcasting
Kathmandu, Nepal

NETHERLANDS

Radio Nederland
P.O. Box 222
Hilversum, Holland

NEW HEBRIDES

Radio Vila
New Hebrides Broadcasting
Service
P.O. Box 110
Vila, New Hebrides

NEW ZEALAND

Radio New Zealand
External Services Division
P.O. Box 2092
Wellington, New Zealand

NIGERIA

Western Nigeria Broadcasting
Service
P.O. Box 1460
Ibadan, Nigeria

NORWAY

Radio Norway
Oslo, Norway

PAKISTAN

Radio Pakistan
82-A, Satellite Town
Rawalpindi, Pakistan

PANAMA

Voice of del Baru
Apt. 160
David, Panama

PAPUA NEW GUINEA

Radio Port Moresby
P.O. Box 1359, Boroko
Port Moresby, Papua

PHILIPPINES

Far East Broadcasting
Company
Box 2041
Manila, Philippines

Voice of the Philippines
National Media Production
Center
Solana Street Intramuros
Manila, Philippines

POLAND

Polskie Radio & Television
Woronicza 17
Warsaw, Poland

PORTUGAL

Portugal National Radio
Rua do Quelhas 21
Lisbon 2, Portugal

RHODESIA

Radio Rhodesia
Rhodesia Broadcasting Corp.
P.O. Box HG444
Highlands, Salisbury,
Rhodesia

ROMANIA

Radio Bucharest
P.O. Box 111
Bucharest, Romania

SABAH

Radio Malaysia, Sabah
P.O. Box 1016
Kota Kinabalu, Sabah

SAO TOME AND PRINCIPE

Radio Sao Tome
Caixa Postal 44
Sao Tome E Principe

SARAWAK

Radio Malaysia, Sarawak
Broadcasting House
Kuching, Malaysia

SAUDI ARABIA

Broadcasting Service of
Kingdom of Saudi Arabia
Ministry of Information
Riyadh, Saudi Arabia

SENEGAL

Senegalese Broadcasting
System
International Service
B.P. 1765
Dakar, Republic of Senegal

SEYCHELLES

Far East Broadcasting
Association
Box 234
Mahe, Seychelles

SIERRA LEONE

Sierra Leone Broadcasting
Service
New England,
Freetown, Sierra Leone

SINGAPORE

Radio Singapore
Department of Broadcasting
Ministry of Culture
P.O. Box 1902, Singapore

SOLOMON ISLANDS

Solomon Islands
Broadcasting Service
P.O. Box 1
Honiara, Solomon Islands

SOMALIA

Radio Mogadiscio
Ministry of Information
& National Guidance
Private Postbag
Mogadiscio, Dem. Rep. of
Somalia

SOUTH AFRICA

Radio RSA
The Voice of South Africa
P.O. Box 4559
Johannesburg, South Africa

SPAIN

Radio Nacional de Espana
Foreign Service
Casa de la Radio
Prado del Rey
Madrid 24, Spain

SRI LANKA

Sri Lanka Broadcasting
Corporation
P.O. Box 574
Colombo 7, Republic of
Sri Lanka

SWEDEN

Radio Sweden
S-10510
Stockholm, Sweden

SWITZERLAND

Radio Switzerland
SBC Overseas Service
Giacomettistrasse 1
CH-3000
Bern 16, Switzerland

SYRIAN ARAB REPUBLIC

Syrian Broadcasting &
Television Service
Place des Ommayades
Damascus, Syrian Arab
Republic

TAHITI

Radio Tahiti
B.P. 125
Papeete, Tahiti

TAIWAN

Voice of Free China
Broadcasting Corp. of China
53 Jen Ai Road, Sec. 3
Taipei, Taiwan

TANZANIA

Radio Tanzania
External Service
P.O. Box 9191
Dar es Salaam, Tanzania

THAILAND

Radio Thailand
Government Public Relations
 Department
Bangkok, Thailand

TOGO

Radio Togo
B.P. 434
Lome, Republic of Togo

TURKEY

Turkish Radio-Television Corp.
Nevzat Tandogan Caddesi 2
Kavaklidere
Ankara, Turkey

UGANDA

Uganda Broadcasting Corp.
Ministry of Information &
 Broadcasting
P.O. Box 2038
Kampala, Uganda

UNITED NATIONS

United Nations Radio Service
New York, New York, USA

USA

Family Radio Network, Inc.
290 Hegenberger Rd.
Oakland, CA 94621, USA

World International
 Broadcasters
P.O. Box 88
Red Lion, PA 17356, USA

American Forces Radio &
 Television Service
Washington, DC 20305, USA

Voice of America
US Information Agency
Washington, DC 20547, USA

USSR

Radio Kiev
Kiev, USSR

Radio Moscow
Moscow, USSR

Radio Vilnius
Vilnius, USSR

VATICAN CITY

Vatican Radio
Vatican City, Vatican

VIETNAM (Democratic Republic of)

The Voice of Vietnam
58 Quan-Su Street
Hanoi, Democratic Republic
of Vietnam

YEMEN ARAB REPUBLIC

Radio San'a
Ministry of Information
San'a, Yemen Arab Republic

YUGOSLAVIA

Radio Belgrade
Chief Editor of External
Broadcasting
2 Hilendarska
Belgrade, Yugoslavia

ZAMBIA

Zambia Broadcasting
Services
Broadcasting House
P.O. Box RW-15
Lusaka, Zambia

Chart 1. SINPO Code

S (Signal Strength)		I (Interference)		N (Noise)		P (Propagation Disturbance)		O (Overall Merit)	
5	Excellent	5	Nil	5	Nil	5	Nil	5	Excellent
4	Good	4	Slight	4	Slight	4	Slight	4	Good
3	Fair	3	Moderate	3	Moderate	3	Moderate	3	Fair
2	Poor	2	Severe	2	Severe	2	Severe	2	Poor
1	Barely audible	1	Extreme	1	Extreme	1	Extreme	1	Unusable

6

Station Log

STATION LOG

LOCATION	FREQ	TIME	REMARKS

STATION LOG

LOCATION	FREQ	TIME	REMARKS

STATION LOG

LOCATION	FREQ	TIME	REMARKS

STATION LOG

LOCATION	FREQ	TIME	REMARKS

STATION LOG

LOCATION	FREQ	TIME	REMARKS

STATION LOG

LOCATION	FREQ	TIME	REMARKS

STATION LOG

LOCATION	FREQ	TIME	REMARKS

STATION LOG

LOCATION	FREQ	TIME	REMARKS

STATION LOG

LOCATION	FREQ	TIME	REMARKS

STATION LOG

LOCATION	FREQ	TIME	REMARKS

STATION LOG

LOCATION	FREQ	TIME	REMARKS